Energy Efficient Cities

Energy Efficient Cities

Assessment Tools and Benchmarking Practices

Edited by
Ranjan K. Bose

ESMAP
Energy Sector Management Assistance Program

THE WORLD BANK
Washington, D.C.

1818 H Street NW
Washington DC 20433
Telephone: 202-473-1000
Internet: www.worldbank.org
E-mail: feedback@worldbank.org

1 2 3 4 13 12 11 10

ISBN: 978-0-8213-8104-5
eISBN: 978-0-8213-8309-4
DOI: 10.1596/978-0-8213-8104-5

Library of Congress Cataloging-in-Publication Data

Urban Research Symposium (5th : 2009 : Marseille, France)

Energy efficient cities : assessment tools and benchmarking practices / edited by Ranjan K. Bose.

p. cm.

Papers presented during the Energy Sector Management Assistance Programme's sessions at the 5th Urban Research Symposium held in Marseille, France from June 28–30, 2009.

ISBN 978-0-8213-8104-5 — ISBN 978-0-8213-8309-4 (electronic)

1. Cities and towns—Energy consumption—Congresses. 2. Energy policy—Congresses. 3. City planning—Congresses. 4. Urban policy—Congresses. I. Bose, Ranjan K. II. Energy Sector Management Assistance Programme. III. World Bank. IV. Title.

HD9502.A2U74 2009
333.79'16091732—dc22

2009053125

Cover photo by Yuri Kozyrev/World Bank.
Cover design by Edelman Design Communications.

Contents

Boxes

Figures

Tables

Foreword

WITH CITIES ACCOUNTING FOR HALF THE WORLD'S POPULATION TODAY, AND two-thirds of global energy demand, urbanization is exacting a serious toll on the environment. As rapid urban growth continues, energy use in cities and associated levels of greenhouse gas (GHG) emissions are projected to continue unabated; current projections indicate that approximately 70 percent of the world's population will live in cities by 2050, producing some 80 percent of the world's GHG emissions. Unfortunately, most of this urban growth will take place in developing countries, where the vast majority of people remain underserved by basic infrastructure service and where city authorities are under-resourced to shift current trajectories. Further, the developing regions of Africa and Asia are where the most rapid urbanization is taking place, and they are least able to cope with the uncertainties and extremities of climate impacts.

The development and mainstreaming of energy-efficient and low-carbon urban pathways that curtail climate impacts without hampering the urban development agenda thus are essential to meeting such challenges. Reducing long-term energy use through efficiency also enhances energy security by decreasing dependence on imported and fossil fuel. In addition, lower energy costs free up a city's resources to improve or expand services while providing important local co-benefits, creating new jobs, enhancing competitiveness, improving air quality and health, and providing a better quality of life.

Energy Efficient Cities: Assessment Tools and Benchmarking Practices has been developed from a careful review of selected papers presented during two ESMAP-sponsored sessions at the fifth World Bank Urban Research Symposium, "Cities and Climate Change: Responding to an Urgent Agenda," which focused on tools and assessment approaches, and on good practices for energy-efficient urbanization. The scope of the papers encapsulates all three urban contexts: new cities, expanding cities, and retrofitting existing cities. The range of policy-relevant conceptual tools and practices discussed during the sessions, and subsequently built upon in this volume, helps achieve a better understanding of leverage points for energy-efficiency interventions and helps catalyze solutions that will delink high levels of carbon-intensive energy use from urban growth without compromising local development priorities.

Jamal Saghir
Director, Energy, Transport, and Water
Chair, Energy Sector Board
Sustainable Development Network
The Word Bank

Acknowledgments

The World Bank's Energy Sector Management Assistance Program (ESMAP) launched the Energy Efficient Cities Initiative (EECI) in October 2008 to help catalyze solutions for reducing energy-intensive urban growth without sacrificing socioeconomic development priorities. This edited volume was developed from a review of selected papers presented during two ESMAP-sponsored EECI sessions at the World Bank's fifth Urban Research Symposium, "Cities and Climate Change: Responding to an Urgent Agenda," held in Marseille, France, June 28–30, 2009. More than 85 participants from 16 countries attended the two ESMAP sessions addressing urban energy issues: one on tools and assessment approaches with regard to energy-efficient urban development, the other on good practices that promote low-carbon pathways.

This volume, edited by Ranjan K. Bose, ESMAP Senior Energy Specialist, World Bank, Washington, DC, with technical support and editing by consultant Sangeeta Nandi, is based on seven multiauthor works. The most important contribution to this edited volume came from the authors, without whose research and efforts to quickly provide complete responses to the intense correspondence that led to the current publication, this book on energy efficient cities would have been a far lesser work. It was a privilege to exchange ideas with these authors and gain their wisdom on city planning methodologies and on tools and good practices that incorporate energy efficiency into the decision making for city energy policies.

This edited volume benefited greatly from peer review by the following experts both within and beyond the World Bank Group: Dilip Ahuja, Arish Dastur, Feng Liu, Sharath Chandra Rao, Matthias Ruth, and Jas Singh. The reviews were constructive and extensive, allowing the authors to reshape the chapters in view of their valuable comments.

Special thanks are due to Sangeeta Nandi, whose diligent reviewer's eye for technical details, together with her editing and writing skills, helped compile the set of papers into a cohesive knowledge product. Special thanks to Amarquaye Armar,

ESMAP Program Manager, for his continuous encouragement and intellectual support; to Heather Austin for editorial support; and to Sharath Chandra Rao for compiling peer review comments and providing logistical support in organizing the two ESMAP sessions in Marseille. Finally, the editor would like to dedicate his efforts, which culminated in this volume, to the memory of his father, the late Chittaranjan Bose, whose memory abides and continues to inspire.

Abbreviations

ABGR	Australian Building Greenhouse Rating
ADB	Asian Development Bank
API	application programming interface
ARCADIA	Adaptation and Resilience in Cities: Analysis and Decision Making Using Integrated Assessment
ASBEC	Australian Sustainable Built Environment Council
BASIX	Building Sustainability Index (used in Sydney and other parts of New South Wales, Australia)
BP	British Petroleum
BREEAM	Building Research Establishment Environmental Assessment Method
BRT	bus rapid transit
CA	city administration
CASBEE	Comprehensive Assessment System for Building Environmental Efficiency
CBD	central business district
CCGT	combined-cycle gas turbine
CDM	clean development mechanism
CEPALCO	Cagayan Electric Power and Light Company
CFL	compact fluorescent lamps
CHED	Commission on Higher Education
CHP	combined heat and power
CIE	Centre for International Economics
COAG	Council of Australian Governments
CTI	computer–telephone integration
DARPTW	Dial-a-Ride Problem with Time Windows
DC	department circular

DECCW	Department of Environment, Climate Change and Water (New South Wales, Australia)
DEWHA	Department of the Environment, Water, Heritage and the Arts (Australia)
DOE	Department of Energy
DRAM	Disaggregate Residential Allocation model
DRT	demand responsive transport
DSM	demand side management
EE&C	Energy Efficiency and Conservation Program
EECI	Energy Efficient Cities Initiative
EELS	energy efficient lighting systems
ELI	Efficient Lighting Initiative
EMPAL	Employment Allocation model
ESCO	Energy Service Company
ESMAP	Energy Sector Management Assistance Program
FCM	Federation of Canadian Municipalities
G2CN	Getting to Carbon Neutral
GBCA	Green Building Council of Australia
GCM	Global Climate model
GDP	gross domestic product
GEF	Global Environment Facility
Gg	gigagrams
GGAS	Greenhouse Gas Abatement Scheme (New South Wales, Australia)
GHG	greenhouse gas (carbon dioxide [CO_2], methane [CH_4], nitrous oxide [N_2O], hydrofluorocarbons [HFCs], perfluorocarbons [PFCs], and sulphur hexafluoride [SF_6])
GIS	Geographic Information Systems
GLA	Greater London Authority
GPRS	General Packet Radio Service
GPS	Global Positioning System
GSM	Global System for Mobile Communications
GWh	gigawatt hour
HDI	Human Development Index
HK-BEAM	Hong Kong Building Environmental Assessment Method
HPS	high pressure sodium
HTS	Household Travel Survey
ICLEI	International Council for Local Environmental Initiatives
ICT	information communication technologies
IEA	International Energy Agency
IEC	Information, Education and Communication Campaign
IG	Implementing Guidelines
IIEC	International Institute for Energy Conservation
IIEE	Institute of Integrated Electrical Engineers

IIEEF	Institute of Integrated Electrical Engineers Foundation
ILUTE	Integrated Land Use, Transportation, Environment modeling system
IPART	Independent Pricing and Regulatory Tribunal (New South Wales, Australia)
IPCC	Intergovernmental Panel on Climate Change
IPP	Investment Priorities Plan
IRR	Implementing Rules and Regulations
ISO	International Organization for Standardization
IT	information technology
ITLUP	Integrated Transportation and Land Use Planning package
ITS	intelligent transportation systems
IVR	interactive voice response
IVT	in-vehicle terminals
kW	kilowatt-hour
LATL	Lighting Appliance Testing Laboratory
LED	light emitting diode
LEED	Leadership in Energy and Design
LILT	Leeds Integrated Land-Use model
MDM	Multisectoral Dynamic model
MEPLAN	input-output based transportation/land-use model
MEPS	Minimum Energy Performance Standards
MILP	mixed-integer linear programming
MOA	memorandum of agreement
MRSFF	Melbourne Region Stocks and Flows Framework
Mtoe	million tonne oil equivalent
MW	megawatt
MUSSA	Land Use Model for Santiago
NABERS	National Australian Built Environment Rating System
NEECP	National Energy Efficiency and Conservation Program
NEPR	National Energy Performance Rating (U.S. Energy Star)
NSW	New South Wales (Australia)
OECD	Organisation for Economic Co-operation and Development
OFWAT	Water Services Regulation Authority (England and Wales)
PCA	Property Council of Australia
PCP	Partners for Climate Protection
PDA	personal digital assistant
PEEP	Philippine Energy Efficiency Project
PELMATP	Philippine Efficient Lighting Market Transformation Project
PLIA	Philippine Lighting Industry Association
PMO	project management office
PNS	Philippine National Standards
PS	procurement service
PUMA	multi-agent modeling of urban systems

RAF	Rapid Assessment Framework
RAMBLAS	regional planning model based on micro-simulation of daily activity travel patterns
RELU-TRAN	Regional Land Use and Transport model
RTN	resource-technology network
SEALS	Scientific Environmental and Analytical Laboratory Services
SMA	Sydney Metropolitan Area
t e CO_2	tonnes of carbon dioxide equivalents
T12	fluorescent lamp (fat tube type) or 1 ½ inches in diameter)
T8	fluorescent lamp (slim tube type or 1 inch in diameter)
TDC	travel dispatch center
TESDA	Technical Education and Skills Development Authority
TRANUS	Integrated Land Use and Transport model
UIAF	Urban Integrated Assessment Facility, Tyndall Centre (United Kingdom)
UK	United Kingdom
UKGBC	United Kingdom Green Building Council
UNFCCC	United Nations Framework Convention on Climate Change
UNDP	United Nations Development Programme
UNEP	United Nations Environment Programme
UN-HABITAT	United Nations Human Settlements Program
UrbanSim	Urban Simulation model
USEPA	U.S. Environmental Protection Agency
VA	voluntary agreements
W	watt
WB	World Bank
whatIf?	Scenario modeling approach developed by WhatIf Technologies Inc. (www.whatiftechnologies.com)

CHAPTER 1

The Imperative of Efficient Energy Use in Cities: Analytical Approaches and Good Practices

Sangeeta Nandi and Ranjan K. Bose

According to current estimates, cities house half of the world's population but account for two-thirds of global energy requirements, with high levels of associated greenhouse gas (GHG) emissions. GHG emissions from urban energy intensity potentially undermine the very development process that energy use and urbanization catalyzes by enhancing the likelihood of anthropogenically generated climate risks at a global scale. Therefore, there is a need for urban management approaches to encompass holistic energy efficiency measures that recognize the environmental implications of carbon-fueled urbanization processes. Against the above backdrop, this chapter provides an introduction to the chapters in this volume. The chapters are based on papers presented at two ESMAP sponsored EECI sessions during the World Bank's fifth Urban Research Symposium, entitled "Cities and Climate Change: Responding to an Urgent Agenda." Categorized under two themes, "tools and assessment approaches on energy efficient urban development" and "good practices that promote low-carbon sector interventions," they variously explore three urban contexts: the formation of new cities; the expansion of old cities; and the retrofitting of existing infrastructure systems and equipment in city sectors. The analytical tools and policy insights discussed extend from integrated assessments of new cities to the impacts of socioeconomic, climate, and demographic changes on existing cities, and offer policy relevant concepts demonstrated through real-world case studies. Sector-specific interventions are deliberated with regard to tools that "green" buildings in Australia, the transformation to efficient lighting systems in the Philippines, and demand responsive transport (DRT) systems in France. In addition, the documentation and benchmarking of a variety of low-carbon and carbon-neutral good practices provide a range of practical insights on plausible energy efficient interventions in urban sectors.

Sangeeta Nandi, PhD, is an Independent Consultant who specializes in sustainable economic development issues; Ranjan K. Bose, PhD, is Senior Energy Specialist at the World Bank.

ENERGY IS INTRINSIC TO URBAN SETTLEMENTS, EMBEDDED IN THE BUILT ENVIRONMENT, directly used to power socioeconomic activity, transport, and communications, as well as to enable the provision of municipal services. According to current estimates, cities[1] house approximately half the world's population but account for two-thirds of global energy requirements, with high levels of associated greenhouse gas (GHG) emissions. It is inevitable that with rapid urban growth, urban energy use will continue unabated: current projections indicate that 70 percent of the world's population will live in urban areas by 2050, with the most rapid urbanization occurring in developing regions (figure 1.1).[2]

Reinforcing statistics on high levels of energy intensity in urban areas, the International Energy Agency (IEA 2008) estimates that global city energy use will grow by 1.9 percent per year, from approximately 7,900 Mtoe (million tonnes oil equivalent) in 2006 to 12,374 Mtoe in 2030, compared to an expected overall global growth rate of 1.6 percent in energy use in the same period. The substantive role of urban processes as a contributing factor to global-climate-threatening GHG emissions was reasserted during a consultative Practitioner's Roundtable Discussion,[3] held October 21–22, 2008, organized by the Energy Sector Management Assistance Program (ESMAP) in cooperation with the Urban Anchor of the World Bank. Designed as a stock-taking exercise, the roundtable concurred that energy efficient urban management is a core policy tool that can acheive significant energy savings. However, the scale of interventions required for this policy in developing country cities is immense because of the complexity of their energy challenge: that of

FIGURE 1.1
Regional Urbanization Trends Worldwide

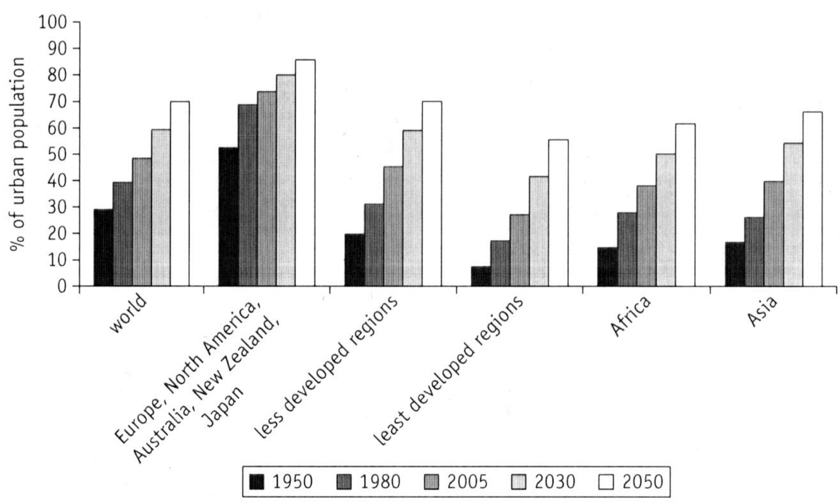

Source: UN 2008a.
Note: UN definitions of less developed and least developed regions. Asia does not include Japan.

implementing energy efficient urban practices while simultaneously ensuring universal access to clean, modern fuels for the underserved urban poor. Achieving this goal calls for committed approaches from both individual governments and the international community, with proactive support from the global development community.

Against this backdrop and in response to the crucial role of urban energy efficiency for environmentally sustainable and inclusive development processes, ESMAP launched the Energy Efficient Cities Initiative (EECI) to facilitate the implementation of practical energy solutions that meet the development priorities of cities and simultaneously build their climate resilience. Energy efficiency through retrofitting existing urban systems is a crucial tool to achieve this goal. Investments to reduce long-term energy use through energy efficient approaches lowers cities' generation of GHG emissions without affecting the scale of urban activity and enhances energy security by lowering dependence on imported fuels. Also, lower energy costs free up resources that can improve or extend city services, in addition to potentially yielding local "co-benefits" such as job creation through higher local investments and improved air quality.

Researching Energy Efficient Urban Solutions: Fifth Urban Research Symposium

ESMAP participated in the World Bank's fifth Urban Research Symposium, "Cities and Climate Change: Responding to an Urgent Agenda," held in Marseille, France, June 28–30, 2009. At the symposium, ESMAP sponsored two sessions, organized by the EECI, titled "Tools and Assessment Approaches" and "Good Practices" for energy efficient cities. This was in consonance with ESMAP's overall objective to mainstream and leverage knowledge and interventions related to urban energy efficiency through EECI. Thus, the themes at the symposium explored energy efficient practices in three urban contexts:

- The formation of new cities.
- The expansion of old cities.
- The retrofitting of existing infrastructure systems and equipment in city sectors.

The development of analytical assessment approaches and tools on urban energy interventions in different urban settings is intended to achieve a better understanding of the various linkages related to urban energy use. This in turn will provide leverage points for energy efficiency interventions and will be ably complemented by EECI outreach on good practices with respect to energy efficiency in urban areas. The overall objective is to systematically bridge the knowledge gap on energy efficient urban solutions between different country groups. In developing countries specifically, data inadequacies often constrain rigorous quantitative analysis of the energy implications of urbanization, either based on past experiences or with respect to future projections. Most of the literature in this regard originates in developed countries and is based on developed country empirics. This research often raises important conceptual issues and solutions that should be explored for their scalability and adaptability in the developing country context. Therefore, it is important to discuss these studies to encourage knowledge dissemination and catalyze their widespread use in decision support systems.

This volume is based on seven topical papers presented at the EECI sessions during the fifth Urban Research Symposium. Chapters 2–8 comprise the papers presented at the two sessions, "Tools and Assessment Approaches," and "Good Practices." Chapter 9 considers the road ahead for ESMAP in its continued support of energy efficiency initiatives in developing country cities. This chapter provides overviews of the papers presented at the symposium with respect to tools and assessments for energy efficient cities and the benchmarking of urban energy efficient good practices. The overviews are presented against a contextual background on the interrelated associations between energy, socioeconomic progress, and urbanization, and the chapter concludes by delving into the energy imperatives of developing countries, which are the focus of ESMAP-sponsored interventions under EECI.

The Background on Energy Use, Socioeconomic Progress, and Urbanization

Two aggregate associations assume importance from the human development perspective in the context of increasing levels of both urban energy intensity and urban population. These are (1) the crosscutting, catalytic role played by energy services in attaining social and economic development objectives, which has been widely documented (for exapmle, DFID 2002; World Bank and UNDP 2005), and (2) the positive relationship between urbanization, a manifestly energy-intensive process, and human development (box 1.1). However, GHG emissions from urban energy intensity potentially undermine the very development process that energy use and urbanization catalyze by enhancing the likelihood of anthropogenically generated climate risks at a global scale. Cities themselves are inherently vulnerable to the risks associated with changing climate, given their high population densities, extensive physical and financial assets, and concentrated entrepreneurial activity within a relatively limited geographical area. As table 1.1 shows, the density of population in urban areas is much higher than average national levels; this is more pronounced in developing regions. Also, as cities grow, disaster risks emanating from both natural and manmade causes often increase through the rising complexity and interdependence of urban infrastructure and services, greater population density, and increased resource concentration (UN-Habitat 2007).

The Intergovernmental Panel on Climate Change (IPCC 2007) estimates of the share of different sectors in total anthropogenic GHG emissions, specifically carbon dioxide, highlight the magnitude of climate impacts directly or indirectly emanating from urban areas. The estimates for 2004 clearly show that urban-dominated sectors like industry, buildings, and transport contribute significantly to GHG

Box 1.1

Urbanization and Human Development Attainments: A Positive Association

Box figure B1.1a plots the relationship between the Human Development Index (HDI), a composite index that measures the average achievements in a country with respect to three basic dimensions of human development—health, access to knowledge, and per capita income (for example, UNDP 2009)—and the percentage share of urban population in 120 countries. These countries range across the spectrum of human development attainments, from very high to low, as defined by the United Nations Development Programme (UNDP). Index values in the dataset accordingly range from 0.968 for Norway to 0.370 for Democratic Republic of Congo. A similar exercise has also been undertaken with respect to the aggregate relationship between per capita income level, represented by GDP per capita in constant 2000 U.S. dollars, and urban population share in the same 120 countries for which data were available (figure B1.1b). It is extremely significant that the correlation coefficient between human development and urbanization, 0.71, is stronger than that between per capita income level and urban population share, calculated to be 0.57; this is apparent in the nature of the clusters in figures B1.1a and B1.1b. The results obtained from this simple exercise appear to reflect the critical importance of urbanization as a development pathway: alongside income-generating economic opportunities, urbanization makes available, within a concentrated spatial area, basic socioeconomic amenities that are vital to the build-up of human capital and future productive capacity. Proximity to basic provisions enables access, especially for the poor, and existing datasets reflect this to an extent in their compilations, which show higher levels of access to safe water and sanitation facilities in urban areas compared to rural areas in the developing world.

FIGURE B1.1a

World Urban Population Share and HDI Value, 2005

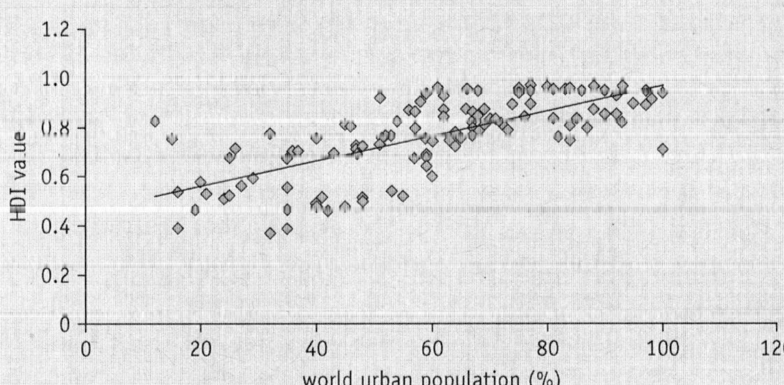

Source: Urban population share: UN 2008a; HDI: UNDP 2009.

(*continued*)

Box 1.1 (*continued*)

FIGURE B1.1b

World Urban Population Share and GDP per Capita, 2005

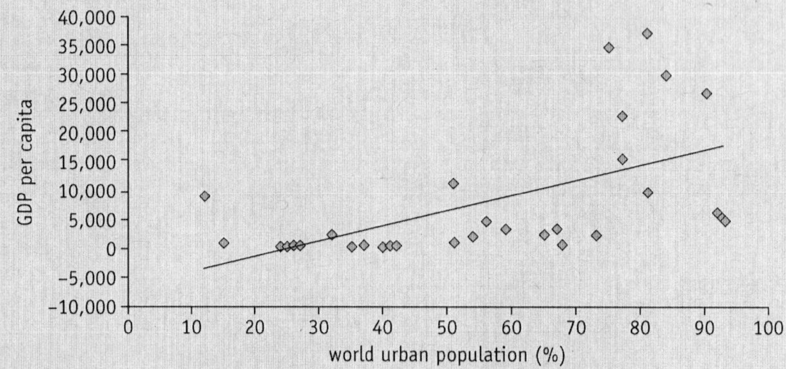

Source: Urban population share: UN 2008a; GDP per capita: World Bank 2009.

TABLE 1.1

World Population Density Levels, 2005

World region	*Average population density (population per km²)*	*Urban population density (population per km² of urban extent)*
More developed regions	23	482
Less developed regions	64	1,381
Least developed regions	37	2,546

Sources: Average population density: UN 2009; Urban population density: UN 2008b.

emissions; this is in addition to the process of supplying energy itself, deforestation, agriculture, and the provision of water and wastewater services (figure 1.2). While disaggregated sector-specific or emission-specific data are not available on GHG emissions from urban systems, IEA (2008) estimates reveal that more than 70 percent of global GHG emissions is currently generated in urban centers (box 1.2).

Environmental stresses originating from the considerable generation of GHGs in cities are of two types: localized air pollution that impacts human health and productivity levels and long-term global climate impacts. The pervading existence of urban poverty in developing countries in Asia and Africa, where most rapid urban growth is projected, implies that large segments of the population have low coping capacities with regard to the uncertainties of global climate change. Therefore, it is

FIGURE 1.2

Share of Different Sectors in Anthropogenic GHG Emissions Worldwide, 2004

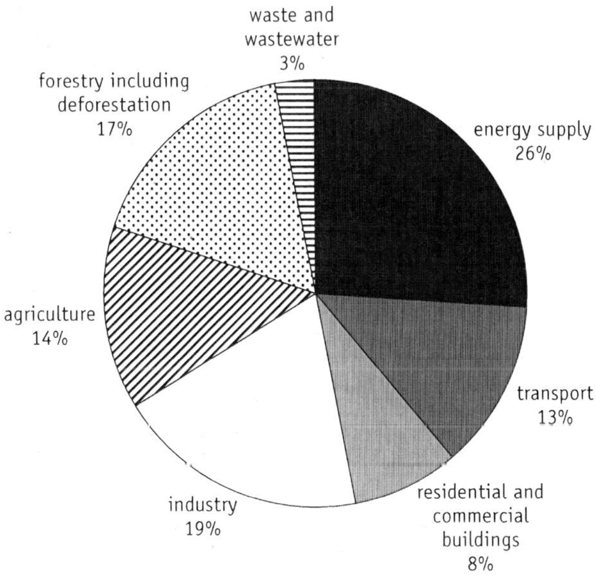

Source: IPCC 2007.

Box 1.2

Urban Energy Use and Associated GHG Emissions

Cities consumed about two-thirds of the world's energy use—an estimated 7,900 Mtoe—in 2006 and accounted for 71% of global GHG emissions, while representing only half the world's population.

By 2030, cities are projected to account for approximately 73% of global energy demand, 76% of global GHG emissions, and 60% of the total global population—equivalent to the total global population in 1986.

Approximately 81% of the estimated increase in energy use in cities between 2006 and 2030 has been projected to derive from non-OECD (Organisation for Economic Co-operation and Development) countries.

Source: IEA 2008.

increasingly important to development that urban energy management approaches shift beyond ensuring continued access to energy sources to encompass holistic energy efficiency measures that recognize the environmental implications of carbon-fueled urbanization processes.

Tools and Assessment Approaches for Energy Efficient Cities

Two broad categories of analytical approaches to identify and prioritize energy efficiency and GHG-reducing interventions in urban areas are (1) integrated platforms that examine systemwide variables to obtain a clear understanding of the multiple dimensions and interlinkages that drive city metabolism and determine urban energy use patterns; and (2) sectorwise energy efficiency intervention strategies. The intervention approaches adopted may comprise both temporal and spatial scales of analysis that factor in the inherent economic and demographic dynamics that characterize urban systems—in their entirety and with respect to specific sectors. These strategies are discussed below with reference to the papers presented at ESMAP's fifth Urban Research Symposium. The papers focus on both integrated, systemwide assessments of new cities and expanding cities and sector-specific energy efficiency retrofits.

New City Tools: Low Carbon Pathways

The energy intensity and associated GHG emissions levels of core urban functions—the provision of socioeconomic opportunities, mobility, and shelter—are largely impacted by the urban form, that is, the physical layout and design of the city, and the economic structure of productive activities within the city. It is generally accepted that compact urban forms are more energy efficient compared to extended urban sprawl, although several factors determine this, including the quality and use of public transport systems and levels of congestion in the city layout. The emerging concept of "eco-cities"—designed to be low carbon or carbon neutral and to limit waste generation to the extent possible (box 1.3)—presents city planners with a unique

Box 1.3

Masdar: A Planned Oil-Free Eco-City in Abu Dhabi, an Oil-Rich Sheikhdom

The city of Masdar in Abu Dhabi, under construction since 2006, is being designed to be oil free. Subterranean electric cars—dubbed "Personalized Rapid Transit"—will ferry passengers from point to point, and solar power plants will provide electricity for lighting and air conditioning and for desalinating ocean water. Wind farms will contribute to the city's energy requirements, along with efforts to tap geothermal energy buried deep beneath the earth. There are also plans to build a plant that will produce hydrogen as well as fuel from residents' sewage in the municipality, which ultimately will aim to be zero carbon and zero waste. Perhaps most important for the desert city, all water will be recycled; even residents' wastewater will be used to grow crops in enclosed, self-sustaining farms that will further recycle their own water.

Source: Biello 2008.

opportunity to use well-calibrated analytical approaches that minimize the carbon footprint of planned cities.

In the above context, chapter 2, "SynCity: An Integrated Tool Kit for Urban Energy Systems Modeling"—based on the paper presented in the first EECI session on "Tools and Assessment Methods" at the symposium—provides a unique method to evaluate holistic urban energy strategies from the early master planning stage for new cities through to assessing the impacts of specific energy supply strategies. Seeking to provide a temporally and spatially diverse representation of urban energy use, but within a generalized framework, the "synthetic city" (SynCity) tool kit model-structure begins with the planning and layout of a new city, incorporating the socioeconomic structure of the city and its activities and the choice of energy carriers and technologies used to meet these demands. At the heart of the model are three submodels, each designed to handle a specific part of the urban energy system analysis: (1) a layout model that seeks to optimize the spatial plan of the city; (2) an agent-activity model that incorporates citizen behavior and generates demand for resources, such as electricity, heat, or transport fuel, and creates resource supplies; and (3) a resource-technology network optimization model that seeks to design the overall energy-supply strategy for the city. In addition, the SynCity model comprises a database to describe and store resources, processes, and technologies, as well as a software program to assemble the model and coordinate running it.

The chapter includes an analysis of the model's application to the prospective development of a new UK eco-town. However, the model has been designed to be flexible for use in existing city contexts subject to a detailed list of adaptations and caveats. An important insight brought out by the application of the model to a prospective UK "eco-town" is in regard to a policy constraint on urban density, brought about by requirements to have adequate green spaces that provide both amenity and ecosystem services in urban centers. In this context, the application of the model to a developing country city would enable a better understanding of the nature and impacts of socioeconomic and policy constraints specific to it on efforts to optimize systemwide efficient energy use. This in turn would yield valuable leverage points for integrated energy efficiency policy initiatives in the city. The authors summarize three major policy uses of the model, which is still at the prototype stage: (1) a simplified version of the tool kit could be used to rapidly assess the major trade-offs within urban energy systems; (2) the optimization model could be used to explore the limits of low-energy urban design, for example, those posed by policy constraints on the density of residential buildings; and (3) the underlying database of urban energy resources and technologies offers a useful reference tool.

Expanding City Needs: Efficient Energy Use to Limit GHG Emissions

Cities expand to accommodate growing economic activity and population. Often city expansion is an organic process, a function of economic stimuli that attract large numbers of people, as well as natural increases in the urban population (that is, increases in the native-born population due to higher birth rates, lower death rates, or both). This pattern is most characteristic of urban growth in developed countries, where physical city expansion typically leads to the incorporation of suburban pockets connected to economic sectors through transport networks. In many developing country cities, on the other hand, rural hardships are responsible for a large proportion of urban population growth, although many migrants from rural areas do not have the skill sets or the economic ability to integrate into the established urban mainstream (Watson 2009). Many settle in informal slums and squatter settlements within the city, as well as in areas peripheral to the city limits, also known as peri-uban areas. Peri-urban areas are situated between consolidated urban regions and the agrarian countryside and, as noted by Satterthwaite (2007), Torres (2008), and others, also host middle- and upper-income residents in addition to low-income households. However, in developing countries, these peri-urban areas tend to be inadequately serviced by municipal provisions, to have poor urban infrastructure, to be characterized by mixed land-use patterns, and to fall in a gray area in the policy context (Torres 2008), although in time they may integrate with the formal city they adjoin. It is natural that assessments of the energy requirements of these two distinct types of urban expansion, suburban and peri-urban, will be determined by spatial features; residential, transport, and other sector requirements; and socioeconomic and demographic aspects that are typical to them. The analytical approaches adopted would necessarily differ too, both methodologically and—depending on the urban relationships and spatial and temporal considerations—in the perspectives that are being studied.

In the developed country context, two integrated city-scale assessment approaches were presented in EECI's first session on "Tools and Assessment Approaches" at the symposium. One paper analyzed the impacts of urban land-use planning policy on residential energy and transport requirements and GHG emissions, and the other assessed climate impacts arising from changes in urban demography, economy, land use, infrastructure, and their interactions. Both analytical tools were developed to inform sustainable urban policy construction in two large cities, London and Sydney, which are important hubs of socioeconomic activity, continually attracting new residents. Thus, they offer policy-relevant conceptual tools that potentially adapt to the analysis of different city contexts, albeit subject to the availability of the required datasets. These analytical approaches are briefly described as follows.

City-Scale Integrated Assessment of Climate Impacts, Adaptation, and Mitigation

Chapter 3, a "City-Scale Integrated Assessment of Climate Impacts, Adaptation, and Mitigation," discusses the model Urban Integrated Assessment Facility (UIAF) being developed by the Tyndall Centre for Climate Change Research, UK. The objective of the model is to analyze long-term change processes in cities and how climate-related drivers interplay with demographic and socioeconomic drivers over timescales of up to 100 years into the future. The chapter discusses the application of the model to London, UIAF's research focus, and its use as an analysis tool to develop energy policy options for the city. The approach is a quantified scenario analysis: a number of modules simulate demographic and socioeconomic changes and their corresponding effects on land use patterns, climate, and GHG emissions over the course of the twenty-first century. The scenarios obtained provide the basis for examining the effect of adaptation and mitigation decisions at the scale of whole urban systems, with particular emphasis on decisions that will have long-term effects. The extended timescale of the analysis coincides with the typical timeframe for assessment of climate change policy and is also motivated by the long life of infrastructure systems and the extended legacy of planning decisions.

The model assembles long-term projections of demography, economy, land use, climate impacts, and GHG emissions within a coherent assessment framework, and it allows the analysis of mitigation and adaptation strategies at a systems scale. It thus provides the flexibility to test a wide range of mitigation and adaptation policies that incorporate land-use planning, modifications to transport systems, changing energy technologies, and measures to reduce climate risks. Significantly, the emissions accounting tool in the model can be used to test the potential effectiveness of energy efficiency measures at varying levels of uptake. In general, the assessment approach adopted shows that it is both possible and informative to look at cities in an integrated and quantified manner over multiple decades.

Integrated Urban Models to Respond to Climate Change in Cities

Using data for Sydney, Australia, as a case study, chapter 4, "Using an Integrated Assessment Model for Urban Development to Respond to Climate Change in Cities," focuses on residential, in-house energy and transport demands, offering a visual support tool to examine how land use planning affects these demands and the associated levels of GHG emissions. The core of the model divides the urban region into separate subregions based on residential location choice by household type; residential location choice is calibrated by population and demographic characteristics and housing type. Submodels are then used to calibrate rates of residential in-house energy and transport use according to household and demographic factors. This process generates a picture of spatially heterogeneous energy consumption patterns across the metropolitan area, enabling an appreciation of factors, such as the distribution of urban infrastructure, that can create considerable variation in energy consumption and related GHG emissions among districts within cities. The energy impacts of policy decisions that affect, by way of example, where new housing is to be built and of what type, can then be simulated. The research offers a policy scenario to monitor progress toward a 2030 vision for a sustainable Sydney. The model can also be configured for interactive scenarios, thus lending itself to use in deliberate city management processes.

An important insight from this analysis relates to energy used for transport purposes in Sydney. Although transport-related emissions dominate the overall spatial distribution of emissions in the city, model results indicate that transport energy use is influenced primarily by proximity to public transport: people, regardless of household structure or income level, use modes of public transport if they live close to them. This points to the important role that public transport plays in mitigating GHG generation in cities. Thus, as indicated by the research results, developing countries should promote efficient, well-connected, comfortable public transport systems to discourage higher private vehicle use as income levels increase, while developed countries should emphasize sustaining and improving the quality of their public transport. However, this may not always be the case: as chapter 4 points out, transport policy in Sydney in recent decades has favored road development over investment in public transport.

Retrofitting and Harnessing Technological Developments in Existing City Sectors

Under its EECI program, ESMAP has identified six city sectors as critical to the implementation of energy efficiency measures: transport, buildings, water/wastewater, public lighting, solid waste, and power/heating. Energy efficiency interventions in these sectors could lead to significant energy and operational cost savings through the replacement of old technologies and devices, and they have a crucial co-benefit in reduced GHG generation (box 1.4). However, resource constraints and lack of access to appropriate technology often act as major barriers to the adoption of energy-saving measures, especially in developing countries. This section provides an overview of three initiatives in the building, lighting, and transport sectors, respectively, which were presented as case studies in the second EECI session on "Good Practices" at the symposium in Marseille on June 29, 2009.

Australian Green Building Rating Tools

Chapter 5, "Green Star and NABERS: Learning from the Australian Experience with Green Building Rating Tools," analyzes tools used to gauge a structure's energy and resource intensity, which can be evaluated on the supply side from two perspectives: building design and building performance. Both types of building assessment systems are used in Australia: Green Star examines the design of the building system, whereas NABERS assesses the actual performance of buildings in energy and resource use over a 12-month period. The chapter discusses the features of the rating tools and their different approaches to assessing a range of categories, including energy, water, and indoor environment, in a building. The chapter also addresses how to achieve increased harmonization between the two types of tools to better realize the potential of ratings tools to "green" buildings. It recommends that green building rating tools be complemented by policy initiatives, such as tax incentives for building owners who retrofit their buildings, to encourage their widespread use.

While acknowledging that widespread use of green building rating tools would be an important step in mitigating climate change, the chapter proposes that building rating tools can also assist in adaptation to climate change by incorporating

Box 1.4

Benefits of Energy Efficiency Improvement in City Sectors: Case Studies

In Fortaleza, Brazil, the local utility implemented measures to improve the distribution of water while reducing operational costs and environmental impacts. With an investment of only US$1.1 million to install an automatic control system and other simple measures, the company has saved US$2.5 million, or 88 GWh over 4 years. More important, the utility was able to establish an additional 88,000 new connections without increasing their overall energy use.

In South Africa, Durban conducted a pilot project to "green" two of its municipal buildings through multiple low-cost measures, resulting in annual energy savings of 15%, or 400,000 kWh, a reduction in CO_2 emissions of 340 tonnes, and generated a return on investment in only 5 months. The city is now reviewing plans to retrofit many more buildings.

In Bogota, Columbia, the implementation of the first two phases of a bus rapid transit (BRT) system, through a combination of advanced Euro II and III technology buses and improved operational efficiencies, has resulted in fuel savings of 47% relative to the city's old public bus network. Additionally, a 32% reduction in overall travel time for commuters, 40% reduction in emissions, and 92% reduction in accident rates in the BRT corridors have been recorded. With the successful registration of Phases II–VIII of its BRT system with the United Nations Framework Convention for Climate Change, the city also expects US$25 million CDM (Clean Development Mechanism) carbon credits by 2012.

Source: ESMAP documentation.

design features that enhance a building's capacity to adapt to weather extremities. To some extent, as the chapter points out, this is already being accomplished through building insulation techniques that reduce energy use while also reducing vulnerability to weather extremes (Ürge-Vorsatz and Metz 2009). In addition, the rating variables can easily be adapted to local climatic conditions and urban planning requirements of different countries. Building rating tools are an important energy efficiency intervention mechanism, given that rapid urban expansion and increasing income levels are inevitably accompanied by large-scale construction activity in cities. Suitable policy incentives can help institute the widespread use of these tools.

Philippine Transition to Energy Efficient Lighting Systems

Chapter 6, "Efficient Lighting Market Transformation in the Making—The Philippine Experience" discusses approaches adopted by a Philippines Department of Energy project to effect the country's transition to an energy efficient lighting economy and documents the achievements of the project. The Philippines Efficient Lighting Market Transformation Project (PELMATP), launched in 2005, is a US$15-million, five-year project, partially supported by a $3.1 million United Nations Development

Programme–Global Environment Facility grant. The chapter details the supply- and demand-side market transformation approaches the project undertook to address barriers to the widespread use of energy efficient lighting systems (EELS). In this context, government bodies and the residential, commercial, and industrial sectors are specifically being encouraged to transition to EELS. The project has targeted an aggregate savings of 29,000 GWh over a decade, equivalent to a nearly 21 percent reduction relative to the business-as-usual Philippines energy efficiency scenario. Two outcomes of the project are the adoption of a national standard for lighting products and the mandatory use of EELS in government offices.

As the chapter notes, the strategy of replacing dated energy-intensive lighting systems with EELS can be easily replicated across sectors and is a "low-hanging fruit" solution to aid energy security and climate mitigation programs. Also, resources freed from lower energy bills can be diverted to other socioeconomic development imperatives, and a strong business case can be presented to city managers and entrepreneurs on long-term cost savings from efficient lighting. The Philippine experience with EELS also highlights the importance of integrated institutional approaches to energy efficiency interventions and the need for both demand-side and supply-side measures.

The Use of Emerging Technologies to Meet a Sector Gap in Services

Chapter 7, "The Role of Intelligent Transport Systems for Demand Responsive Transport," explains the innovative use of information communication technologies (ICT) in the provision of "demand responsive transport" (DRT) systems. Lying between public transport and private transport in the continuum of transport services, DRTs are important because they meet service gaps or complement existing transport systems. Specifically, they cater to dispersed mobility needs, providing transport services when the provision of conventional public transport would be too expensive, for example, during hours of low demand, in areas with low population density, and for target users like the elderly and students. DRT services have been steadily growing in France and other European countries, and the effective harnessing of ICT by DRT systems has been key to their growth. As two case studies in the chapter show, ICT has allowed DRT service structures to become increasingly flexible in their response to passenger mobility needs, which positively impacts the demand for them. A significant shift from the use of private transport to DRT services by regular commuters also offers potential for energy savings.

The DRT concept offers an interesting example of innovative use of modern technological developments to meet a sector gap. DRT services can be an effective means of improving intermodality and system integration, both in overburdened city transport systems as well as in areas inadequately served by public transport, for example, distant suburbs. Their adaptability to local needs is a significant factor in helping meet complex and continually increasing urban mobility needs. However, the higher cost of DRT services compared to public transport may impact their use as a transport alternative in developing countries, although urban areas provide opportunities to explore economies of scale that will lessen the cost of DRT operations.

Benchmarking Good Practices for Achieving Energy Efficiency in Cities

The documentation of good practices in a structured manner is vital to energy efficient urban management as the sharing of successful experiences constitutes an important learning platform for city managers. The variables used to analyze the efficacy of energy efficiency interventions in urban sectors include the cost-effectiveness of the project, the level of CO_2 savings from the project, and the potential returns on project investment. The approach adopted depends on several factors, the most important being data availability and the objective of the intervention. In this regard, efforts are under way to arrive at a clear set of criteria to govern the collation and evaluation of information on energy efficient urban initiatives across sectors and countries by the International Energy Agency (Jollands, Kenihan, and Wescott 2008) and other prominent organizations. These organizations include the UN-Habitat, the United Nations Environment Programme, the Organisation for Economic Co-operation and Development, the C40 Large Cities Climate Leadership Group, the International Council for Local Environmental Initiatives, and more recently, the Energy Sector Management Assistance Program, as well as the Urban Anchor of the World Bank.

Presented in the second EECI session on "Good Practices" at the fifth Urban Research Symposium, chapter 8 provides a review of best practices with regard to urban infrastructure strategy, based on research that informed a guidebook for Canadian municipalities on approaching carbon neutrality. The chapter, "Getting to Carbon Neutral: A Review of Best Practices in Infrastructure Strategy," examines the performance of energy efficiency and GHG-reducing urban initiatives based on their cost-effectiveness in reducing GHG emissions. The case study approach adopted for the assessment entailed the initial collation of data on approximately 70 good practices in carbon-neutral urban design, with the availability of relevant data determining the subsequent benchmarking of the initiatives. The cost-effectiveness of the case studies was found to vary between 3 and 2,780 t e CO_2/yr/US$ million (tonnes of carbon dioxide equivalent per year per million U.S. dollars), where cost-effectiveness is defined as annual GHG emissions saved per dollar of capital investment. A comparative analysis with other datasets indicates that the average cost-effectiveness of the projects in the database of 550 t e CO_2/yr/US$ million is significantly exceeded by solid waste projects in Canada as well as by projects based in developing countries under the Clean Development Mechanism.[4] The chapter advocates reducing greenhouse gases by at least two orders of magnitude in developed country cities in order to have a meaningful global impact on climate change.

Chapter 8 emphasizes the importance of structured databases with respect to all urban infrastructure sectors as a prerequisite for the benchmarking and sharing of good practices. The chapter also significantly contributes to the literature by documenting many types of low-carbon and carbon-neutral urban energy initiatives. In addition, the case studies analyzed could be taken forward to explore their potential for adaptability and scalability in the developing country context. This is especially relevant with regard to urban sectors such as solid waste, assessed as "low-hanging fruit" for cost-effective GHG reduction, since these sectors are underserviced in most developing country cities. The chapter also notes that a diverse range of effective strategies can be adopted to reduce GHG emissions emanating from urban activity; these efforts can exploit favorable local conditions or can be undertaken in response to local environmental stresses.

The Energy Efficiency Imperative in Developing Country Cities

The strong positive correlation between human development attainments and urbanization underlines the importance of opportunities cities offer for socioeconomic progress. However, given the high levels of energy intensity associated with urban systems, the challenge is to gradually delink energy intensity from urbanization. This challenge is magnified in developing regions, where the growth rate of the urban population is much higher than in the more developed, high-income countries (figure 1.3) and the population continues to be underserved in its access to basic amenities like housing, water, and sanitation. In addition, access to modern fuels may also be uncertain for the urban poor. Thus, absolute increases in energy requirements are a development imperative to enhance standards of living in developing country cities. This makes critical the systemwide implementation of energy efficient policies and initiatives that will curtail climate impacts originating in urban energy use without hampering the urban development agenda.

It is important to tailor energy efficiency and climate mitigating interventions to specific needs of cities because each urban settlement is characterized by different

FIGURE 1.3

Annual Growth Rate of Worldwide Urban Population

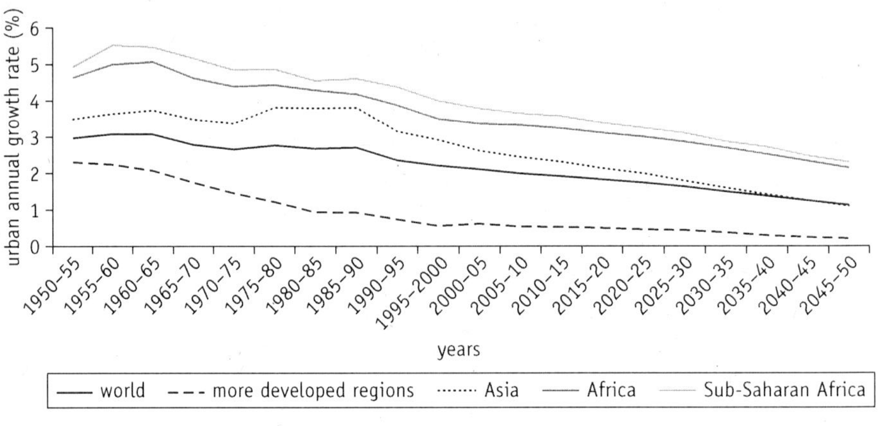

Source: UN 2008a.

socioeconomic, demographic, climate, and topographical patterns. Also, cities and towns in developing countries are at different levels of urban transition, thus encountering varying policy challenges, each associated with specific energy requirements and environmental complexities. In fact, it has been estimated that 51.8 percent of the world's urban population lives in urban centers with fewer than half a million residents (UN 2010); many of these small urban centers can be defined interchangeably as "large villages" or "small towns" (Satterthwaite 2006). Thus, it is vital that assessment of the development, infrastructure, and energy requirements of small urban settlements take into account their socioeconomic vulnerabilities. This consideration also applies to peri-urban settlements, which, as mentioned earlier, tend to integrate over time with the cities that they adjoin.

In conclusion, the fostering of holistic energy efficient urban solutions is urgently required in a rapidly urbanizing global context marked by differing levels of socioeconomic development, intensive energy use patterns, and long-term climate threatening GHG emissions. However, the formulation of energy solutions within developing countries that incorporate their specific characteristics is hindered greatly by inadequacies in their existing information bases. Additionally, as noted earlier, pervading data constraints typically impact the adaptability and scalability of tools, assessment approaches, and policy initiatives emerging from developed countries on energy efficient and climate mitigating interventions. In this context, the creation of a systematic body of energy- and climate-related information in developing countries will aid in knowledge generation and outreach, both immensely important for a cohesive approach to instituting, monitoring, and evaluating energy efficient city initiatives. An additional co-benefit would be the informed participation of stakeholders in the process of energy efficient retrofitting; this can potentially act as a catalyst for innovative and organic urban energy solutions that emerge in response to local conditions and requirements. Thus, the need to mainstream and disseminate knowledge on urban energy efficiency in developing countries also offers an opportunity to city authorities to ensure data collation and availability, thereby facilitating the development of wide-ranging urban energy solutions. Many of these solutions may be rooted in the dynamism inherent in the workings of the urban environment itself.

Notes

1. Following *World Energy Outlook* (IEA 2008), "cities" refers to all urban areas including towns; the terms "city" and "urban" are used interchangeably herein.

2. Urban population estimates cited in the text and figures of this chapter are from UN 2008a. "The World Urbanization Prospects, 2009 Revision Highlights," published in March 2010, notes that world urban population currently comprises 51.5 percent of world population; it projects that 69 percent of world population will be urban in 2050, and that cities with fewer than 500,000 people accounted for 51.8 percent of the urban population in 2009 (UN 2010).

3. Proceedings of the roundtable discussion are available at http://www.esmap.org/filez/pubs/1113200823519_EECI_WorkshopProceedings_FINAL.pdf

4. The Clean Development Mechanism (CDM) is part of the broad framework established by the Kyoto Protocol of the United Nations Framework Convention on Climate Change to curtail or reverse the growth of GHG emissions in industrialized countries while simultaneously promoting the attainment of sustainable development objectives in developing countries. The CDM arrangement allows industrialized countries to invest in emissions reduction projects in developing countries as alternatives to more expensive strategies in their own countries.

References

Biello, David. 2008. "Eco-cities: Urban Planning for the Future." *Scientific American*, October 2008 special edition. http://www.scientificamerican.com/article.cfm?id=eco-cities-urban-planning.

DFID (Department for International Development). 2002. *Energy for the Poor: Underpinning the Millennium Development Goals.* http://www.dfid.gov.uk/Documents/publications/energyforthepoor.pdf.

IEA (International Energy Agency). 2008. *World Energy Outlook.* Paris: International Energy Agency.

IPCC (Intergovernmental Panel on Climate Change). 2007. "Fourth Assessment Report (AR4), Climate Change 2007: Synthesis Report." http://www.ipcc.ch/pdf/assessment-report/ar4/syr/ar4_syr.pdf.

Jollands, Nigel, Stephen Kenihan, and Wayne Wescott. 2008. *Promoting Energy Efficiency Best Practices in Cities: A Pilot Study.* Paris: International Energy Agency.

Satterthwaite, David. 2006. "Outside the Large Cities: The Demographic Importance of Small Urban Centres and Large Villages in Africa, Asia and Latin America." Human Settlements Discussion Paper, Urban Change 3, International Institute of Environment and Development (IIED), London. http://www.iied.org/pubs/display.php?o=10537IIED.

———. 2007. "The Transition to a Predominantly Urban World and Its Underpinnings." Human Settlements Discussion Paper, Urban Change 4, International Institute of Environment and Development, London. http://www.iied.org/pubs/display.php?o=10550IIED

Torres, Haroldo da Gama. 2008. "Social and Environmental Aspects of Peri-urban Growth in Latin American Mega-cities." Paper presented at the United Nations Expert Group Meeting on Population Distribution, Urbanization, Internal Migration and Development, United Nations Secretariat, New York, January 21–23. http://www.un.org/esa/population/meetings/EGM_PopDist/P10_Torres. pdf.

UN (United Nations). 2008a. World Urbanization Prospects: The 2007 Revision Population Database. http://esa.un.org/unup/.

———. 2008b. "Urban Population, Development and the Environment 2007." Wall chart. http://www.un.org/esa/population/publications/2007_PopDevt/Urban_2007.pdf.

———. 2009. World Population Prospects: The 2008 Revision Population Database. http://esa.un.org/unpp/.

———. 2010. "World Urbanization Prospects: The 2009 Revision Highlights." http://esa.un.org/undp/wup/Documents/WUP2009_Highlights_Final.pdf.

UNDP (United Nations Development Programme). 2009. "Human Development Report 2009: Overcoming Barriers: Human Mobility and Development." http://hdr.undp.org/en/media/HDR_2009_EN_Complete.pdf.

UN-Habitat 2007. *Global Report on Human Settlements 2007: Enhancing Urban Safety and Security.* Nairobi, Kenya: UN-Habitat. http://www.unhabitat.org/content.asp?typeid=19&catid=555&cid=5359.

Ürge-Vorsatz, Diana., and Bert Metz. 2009. "Energy Efficiency: How Far Does It Get Us in Controlling Climate Change?" *Energy Efficiency* 2: 87–94.

Watson, Vanessa. 2009. "The Planned City Sweeps the Poor Away . . .: Urban Planning and 21st Century Urbanization." *Progress in Planning* 72 (3): 151–93.

World Bank. 2009. World Development Indicators Online. World Bank, Washington DC. http://web.worldbank.org/WBSITE/EXTERNAL/DATASTATISTICS/0,,contentMDK:20398986~menuPK:64133163~pagePK:64133150~piPK:64133175~theSitePK:239419,00.html.

World Bank and UNDP. 2005. *Energy Services for the Millennium Development Goals.* http://www.unmillenniumproject.org/documents/MP_Energy_Low_Res.pdf.

CHAPTER 2

SynCity: An Integrated Tool Kit for Urban Energy Systems Modeling

James Keirstead, Nouri Samsatli, and Nilay Shah

This chapter demonstrates a new tool for the integrated modeling of urban energy systems. Energy is vital to the delivery of urban services, and its role can be considered at many stages in the urban design process. This chapter begins with the planning and layout of a new city and goes through to the socioeconomic structure of the city and its activities and the choice of energy carriers and technologies used to meet the city's energy demands. Unfortunately, existing modeling technologies typically focus on only one of these components and are customized to a single problem context. Therefore, we have developed a "synthetic city" tool kit (SynCity) to facilitate the integrated modeling of urban energy systems across all of these design steps and in a variety of problem environments. After outlining the components of this methodology, we demonstrate how it can be applied to the case of a United Kingdom "eco-town." The discussion then considers the applicability of the SynCity methodology to developing country contexts and highlights potential use cases, policy applications, and deliverables.

Introduction and Objective

In recent years, there has been a surge of interest in urban climate and energy issues. These trends were summarized in two notable reports. First, the UN's most recent population estimates show that over 50 percent of the world's population now lives in urban areas, a number expected to continue rising, particularly in developing nations (UN 2008). Second, the 2008 *World Energy Outlook* (IEA 2008) explicitly focused on urban energy use, noting that approximately two-thirds of the world's primary energy is consumed by cities and again forecasting continued growth in both

Dr. James Keirstead, Research Fellow, Chemical Engineering; Dr. Nouri Samsatli, Research Associate, Chemical Engineering; and Prof. Nilay Shah, Process Systems Engineering; Imperial College London, United Kingdom.The authors gratefully acknowledge the financial support of BP for the Urban Energy Systems project at Imperial College London.

developed and developing countries. These trends—combined with the activities of organizations such as the Global Carbon Project's Urban and Regional Carbon Management theme, the International Council for Local Environmental Initiatives (ICLEI) Cities for Climate Protection, the C40 Climate Leadership Group, Energie-Cités, Columbia University's Urban Climate Change Research Network, and many others—reflect a growing recognition that energy use in cities is a key element in the fight against global climate change.

Unfortunately, the scale of the challenge seems only to be growing larger. In March 2009, the International Scientific Congress on Climate Change in Copenhagen observed that, "given high rates of observed emissions, the worst-case IPCC (Intergovernmental Panel on Climate Change) scenario trajectories (or even worse) are being realized," therefore increasing the likelihood of "abrupt or irreversible climatic shifts" and associated social and economic disruptions (ISCCC 2009). Furthermore, the global economic downturn has restricted credit for new innovations, created a more risk-averse investment climate, and strained public-sector finances. With scarce resources and the need to achieve greater carbon savings, improvements to urban energy systems will increasingly require integrated solutions that deliver maximum economic efficiency and multiple benefits.

Yet, before cities commit to significant new policy initiatives or infrastructure investments, there is a need to improve our understanding of urban energy use and how it might be changed. Typically, this is a job for mathematical modeling or computer simulations. While such models can never provide a definitive answer to policy makers (because modeling technologies are always partially incomplete and policy debates need to consider other forms of knowledge alongside numerical analyses), they are an important contribution to urban sustainability debates (Tweed and Jones 2000).

Urban modeling technologies have been developed to cover a range of applications, but they are most commonly found in the area of land use and activity location. Batty (2007) suggests three major model classes: land use and transportation models (for example, CUSPA 2009); urban dynamics models (for example, aggregate system dynamics); and micro-simulation models (for example, cellular automata and agent-based models). These tools focus primarily on economic or spatial planning issues, but some groups have adapted these technologies to perform climate risk or environmental modeling; a notable example is the UK Tyndall Centre's integrated assessment model, which is currently being applied to the climate risks of London's spatial plan (Dawson et al. 2009).

There have also been efforts to develop tools for urban energy systems modeling. These include a tool for the assessment of energy, water, and waste consumption at a building or small neighborhood level (Robinson et al. 2007); a geographic information systems (GIS) tool for estimating the spatial pattern of energy requirements in an urban area (Girardin et al. 2008); a model assessing the interactions of heat demand and locally available heat sources such as lakes or incinerators (Mori, Kikegawa, and Uchida 2007); and a model combining demand estimation with an energy-management optimization module (Brownsword et al. 2005). While these applications are quite diverse, they demonstrate two important features of existing urban energy system models. First, such models must include a representation of the spatial and temporal variation of urban energy demand. This can lead to significant input requirements, for example, in the form of GIS data or building design information specifications; however, building up the urban system from individual components allows the aggregate effects of small changes to be assessed more readily. Second, these models explore both the supply and demand sides of urban energy use, for example, by optimizing provision strategies. However, in addition to these positive characteristics, these examples show that current practice consists largely of detailed models built for the assessment of a single aspect of existing systems (for example, domestic sector demands in UK households, heat demand in Geneva, and so on). This means that these tools have limited applicability beyond the specific problem case, thus increasing the resources required for data collection and validation in new contexts. Furthermore, they are unable to offer a truly integrated perspective on urban energy use across all sectors and stages of the design process.

The goal of this chapter is to demonstrate an improved technique for the modeling of urban energy systems. This "synthetic city" approach (named SynCity) seeks to provide an integrated, spatially and temporally diverse representation of urban energy use, but within a generalized framework. By dividing urban energy use into a series of separate but integrated models, and through the use of mathematical modeling techniques rather than detailed datasets, our objective is to develop an urban energy modeling tool that is highly portable and adaptable, thus providing information to decision makers in a variety of problem contexts. After first giving an overview of the system (Methodology and Data section), we apply SynCity to the case of a UK eco-town to demonstrate each major component (Analysis and Results section). In the concluding discussion, we consider the limitations of this technique and outline recommendations for its use.

Methodology and Data: The SynCity Tool Kit

The SynCity tool kit is the main activity of the BP Urban Energy Systems project at Imperial College London. Begun in 2005, the interdisciplinary project seeks to "identify the benefits of a systematic, integrated approach to the design and operation of urban energy systems, with a view to at least halving the energy intensity of cities" (Shah et al. 2006). To achieve this goal, the project employs researchers with a range of expertise, including process systems optimization, urban and industrial ecology, transport and land use modeling, energy systems modeling, energy policy, and business strategy.

After reviewing the relevant literature in these fields, the team began to develop SynCity as an integrative test bed for research activities. It was envisioned that the software would enable each team member to perform analyses specific to his or her field of interest, while transparently drawing on the contributions of other team members. The goal, therefore, was to build a system that did not require excessive data collection or parameterization for each model run and that could simulate a variety of urban energy modeling problems. It was in this context that the structure of the SynCity system was developed.

SynCity Structure

The SynCity tool kit consists of three major components: a series of models designed to handle specific urban energy design problems, a unifying ontology and database to describe and store core data objects, and an executive to assemble and coordinate the running of modeling scenarios. These relationships are depicted in figure 2.1 and described in detail below.

FIGURE 2.1
Overview of the SynCity Tool Kit

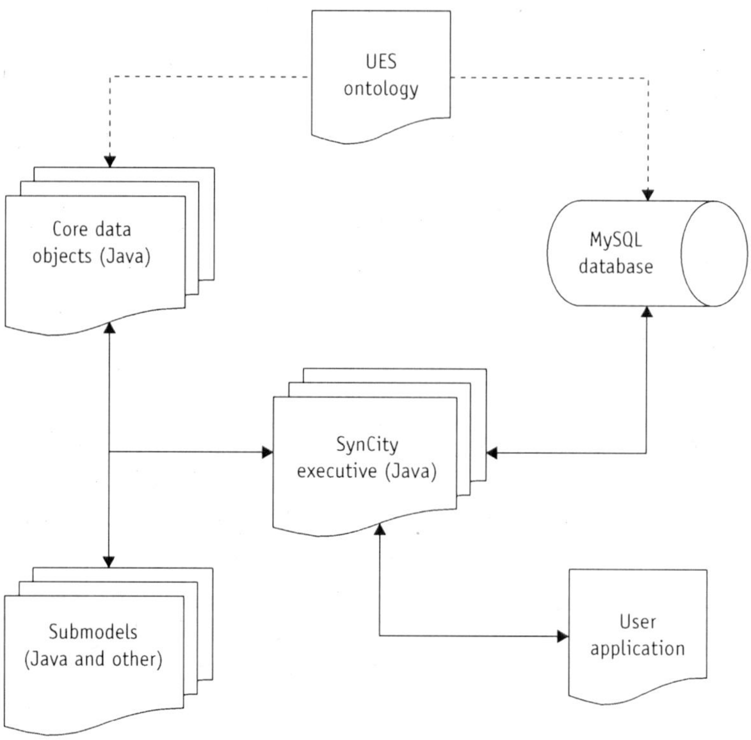

Source: Authors.
Note: The top three components (the urban energy systems ontology, MySQL database, and Java core objects) define the building blocks of an urban energy system, such as resources, processes, technologies, and geography. The user (bottom right) then assembles these components into a model scenario via the SynCity executive. The executive builds, schedules, and executes the individual submodels (for example, the layout, agent-activity, and resource-technology-network models described in this chapter).

The urban energy systems (UES) ontology provides a description of the major objects within an urban energy system. The idea is that this data model can be consistently referenced by both software objects and team members to promote a common understanding of complex concepts. For example, a transport modeler and an electricity system modeler would both share a common view of a network as a mathematical graph, but they might also require additional information about the properties of edges and nodes (for example, electrical resistance, maximum number of vehicles per hour). Five major object categories were identified as follows: (1) *resources*, that is, materials that are consumed, produced, or interconverted including gas, electricity, and so on; (2) *processes*, that is, technologies that convert one set of resources into another set (for example, a gas turbine that converts gas to electricity and waste heat); (3) *technologies*, that is, the infrastructure of a city, including buildings and networks; (4) *spaces*, that is, the physical space of the city and its surrounding hinterlands; and (5) *agents*, that is, the occupants of the city, including citizens, firms, and government actors. These concepts were then codified into an ontology using Protégé (Noy and McGuinness 2002) and converted into a database model and relevant object classes (respectively using the popular open-source MySQL relational database system and Java object-oriented programming language).

The SynCity executive is a Java API (application programming interface) that enables users to manipulate these core data objects and assemble them into modeling scenarios (box 2.1). The API comprises three major class types: data, or object, classes (as described above); model classes for implementing the submodels, either directly in Java or via other software packages; and manager, or utility, classes, which handle file management, model execution, and so on. At present, users build their simulations by writing Java code using this API; however, a graphic-user interface can be built on top of this platform enabling more intuitive model construction. Box 2.1 contains a pseudo-code listing that demonstrates how the API is used in a typical workflow.

SynCity Submodels

The heart of the SynCity tool kit contains three submodels, each designed to handle a specific part of the overall urban energy system analysis. The models are designed to be run sequentially; however, the system is modular and steps can be skipped if not required. Similarly, there are plans to allow the models to feedback within the system, thus facilitating iterative design processes.

Layout Model

The first tool kit component is a layout model designed to optimize the spatial plan of the city. The inspiration for the model comes from flow-based factory design models and optimized urban layout sketches (for example, Feng and Lin 1999; Urban, Chiang, and Russell 2000). As input, the model takes the geography of the city (that is, the size and spacing of each land parcel), the available types of

Box 2.1

Pseudo-code of a Typical SynCity Model

The SynCity tool kit was built using the Java programming language. The pseudo-code below provides an illustration of the workflow for a typical simulation run. After first initializing the system (for example, connecting to the database and preparing working directories), the user creates an empty city and, from the database, loads the processes and resources he or she wishes to have available for subsequent model runs. As an example, for a biomass case study, the user might require woody biomass, electricity, and heat resources of various quality, alongside the processes and technologies needed to interconvert them (for example, a biomass boiler, district heat pipes, electricity networks, and so on). Next, the user initializes the submodels to be run and adjusts any parameters, such as changing the objective function for optimization models or adding custom constraints. In the final step, the user passes the newly constructed city and submodels to a manager object, which then schedules and executes the submodels and reports on the results.

```
// connect to database and other initialization
init();

// Create a manager to handle simulation and store results
Manager m = new Manager("d:/syncity/output /");

// Create an empty city
City c = new City();

// Load relevant data objects from database
c.addResource("elec");
c.addProcess("ccgt");

. . .

// Create submodels
Model layout = new LayoutModel();
Model abm = new TransportABM();
Model rtn = new RTNModel();

// Assign city to manager m
m.setCity(c);

// Add models to manager
m.addModels(layout, abm, rtn);

// Execute models
m.executeModels();

// Results are saved to the working directory
```

residential and commercial buildings, available transportation infrastructures and modes, and average activity profiles of the citizens (for example, shopping, education, employment). Using mixed-integer linear programming techniques, the model then seeks to position buildings, activity locations, and transport networks, subject to a number of constraints. For example, there must be sufficient housing and activity provision for the entire population, each activity can only be provided in certain types of buildings, and buildings can only be constructed on suitable land types. The decision variables of the model are the presence or absence of a building or activity type in a specific cell and the presence or absence of a transportation network link between a pair of cells. The objective function then describes the total "cost" (capital, operating, energy, or carbon) of the resulting infrastructure configuration.

The layout model is typically run for "greenfield" (previously undeveloped) sites and, therefore, has significant freedom in deciding where to locate buildings and networks. However, the model can also be used for "brownfield" scenarios, that is, where urban land is being redeveloped. Where the spatial plan is completely fixed, the user can manually specify these land uses and skip directly to the agent-activity model. Alternatively, if only some land uses are known (for example, those of cells bordering a regeneration project), then the user can add these partial constraints and let the model work within the remaining degrees of freedom. A user similarly can specify the city's relationship to its hinterland. For example, if education must be provided to a new development but neighboring cities already have existing schools, the model can be configured to account for the estimated number of daily visits that these external activities would provide. (The agent-activity model will later simulate the detailed decision of whether an individual attends a local or remote school.)

Agent-Activity Model

The agent-activity model is an agent-based microsimulation model. Beginning with the layout of the city (either manually specified or calculated by the layout model), the model simulates the daily activities of each citizen within the city. This is currently implemented as a simple four-stage transportation model: that is, in each time step of the model, citizens select an activity (*trip generation*), an activity provider (*trip distribution*), a transport mode (*mode choice*), and a route (*trip assignment*). Each of these decisions is stochastic; for example, the trip distribution step uses a gravity-model to assess the attractiveness of alternative service providers. This variability, combined with the interactions between agents (for example, congestion of a network path), allows the agent-activity model to capture some of the complex emergent properties of urban systems.

As the agents within the model perform their daily schedules, demands are created for resources such as electricity, heat, or transport fuel; similarly, resource supplies can be created by agent activities (such as waste) or they can occur naturally (such as ground-source heat or biomass). These supplies and demands are distributed in time and space, providing the primary input for the next model stage.

Resource-Technology Network Model

The resource-technology-network (RTN) model is an optimization model that designs the overall energy-supply strategy for the city. It takes as input the spatially and temporally distributed resource demands determined by the agent-activity model, as well as sets of available process types. These processes describe how resources can be *produced* (for example, converting from gas to electricity), *transported* within the city (to accommodate spatial variance in demand), or *stored* (to accommodate temporal variance in demand). The model's primary constraint is a resource balance, which ensures that resource demands in every cell are satisfied using the outputs of production, transport, and storage processes and importing and exporting resources to and from the surrounding hinterlands. The structure of the RTN model means that it can simultaneously evaluate a number of alternative energy systems. For example, domestic heat demand could be met using gas from mains and household boilers or by using a district heating system, which has a heat network and heat exchangers to move excess process heat from source to demand. More complicated resource chains, for example those involving waste or biomass to energy, can also be modeled within this framework.

The objective function currently minimizes the cost of providing the entire energy supply system (that is, generating equipment, transportation networks, and resource imports). The decision variables describe the location of conversion processes, their production rates, and the rates of resource import and export. The output of the model, therefore, is a map giving the location of each production and storage process and the structure of the resource distribution networks, as well as details of the operational rates of resource import, export, production, and consumption at each time step.

It should be noted that a fourth submodel component, the service network model, is currently in development. This model will convert the macro-scale network designs produced by the RTN model into a more detailed engineering specification.

SynCity Data Requirements

As noted, one of the goals of the SynCity tool kit is to create a model that can be transferred between different problems with minimum additional data collection. At present, the system is backed by a database of approximately 30 resources, processes, and technologies whose parameters are effectively universal (for example, physical properties of resources or efficiencies of technologies). However, the cost data associated with each object is likely to change between locations, and UK price data have been used to date. UK data have also provided the baseline assumptions for the agent-activity model; in other cultures, there may be different preferences for modes of transport or activity locations, and these would need to be taken into account. Nevertheless, we have found that new case studies can be quickly run using the information commonly available on a site master plan (for example, spatial layout, population, and housing densities); this provides an initial result that can then be refined later in consultation with the client.

Analysis and Results: Applying SynCity to a UK Eco-town

A number of potential use cases for the SynCity tool kit have been identified. These include developing the layout of a new development, assessing centralized and decentralized energy provision strategies for a city, and evaluating the potential of refurbishment programs for an existing city. Here we demonstrate the SynCity platform using a proposed UK eco-town as it touches on each component of the tool kit.

Introduction to the UK Eco-Town Case

Given rising demand for housing as well as substantial questions about how the building sector might contribute to national climate change and energy policy goals, the UK government has promoted "eco-towns"—new urban areas of at least 5,000 homes that are exemplar green developments—as an opportunity to drive innovation and to demonstrate how these policy goals might be jointly achieved. While the formal requirements for eco-town certification have not yet been announced by the government, it has been suggested that the headline targets for these developments should be an 80 percent reduction in carbon dioxide (CO_2) emissions (from 1990 levels) and an ecological footprint that is two-thirds of the national average (CABE and BioRegional 2008). Twelve eco-town developments have been put forward for consideration; this chapter considers one of these proposals.

Our analysis focused on one of the design phases of the eco-town site, an area of approximately 90 hectares in central England intended to house 6,500 people. An initial assessment of the proposal by government-commissioned consultants found that the site "might be a suitable location subject to meeting specific planning and design objectives" (DCLG 2008), but more information was required, particularly on the energy strategy for the site. Since then, a study of alternative energy systems has been commissioned by the developers to address some of these criticisms.

The present analysis was designed to assess the performance SynCity platform across each of the three modeling components. The research questions therefore are as follows:

- How do the designs created by the layout optimization model compare with the proposed eco-town master plan?
- Does the current agent-activity model provide a reasonable estimation of energy demands when compared with the design assumptions of the eco-town developer's energy strategy?
- How does the RTN model respond to changing assumptions about the provision of energy services for the eco-town?

Layout Model Results

The developers of the proposed eco-town had already created a master plan for the site, and we were asked to use SynCity to assess this design and explore alternative layouts. The layout model can provide an initial evaluation of such designs by calculating the estimated annual cost (capital and operating), energy and carbon emissions for buildings, and anticipated transportation flows.

We began by converting the geography of the site from a GIS shape file to a simple cellular representation, maintaining the area and relative separation of each cell. Using the master plan, the location of known housing and activity types were fixed before allowing the layout model to estimate the associated transportation flows, as shown in figure 2.2. This step provides a baseline result for comparing alternative designs.[1]

In these plots, each cell is colored according to the building type located on the cell; for example, schools are yellow, parks are green, and housing is blue (with darker shades representing higher dwelling densities). The activity performed at each cell is indicated by the label, and traffic flows between cells are represented by the width of the connecting lines. In all cases, a list of possible network connections is given as input to the model in order to prevent the connection of non-neighboring or otherwise separated cells. Single nodes are also used to collect transport demand from local cells to a specific point, an approach commonly used in transport modeling.

After establishing the baseline master plan layout, the model was run in an "unconstrained" mode, that is, the model sought to provide sufficient housing and

FIGURE 2.2
Layout of the Eco-Town as Planned

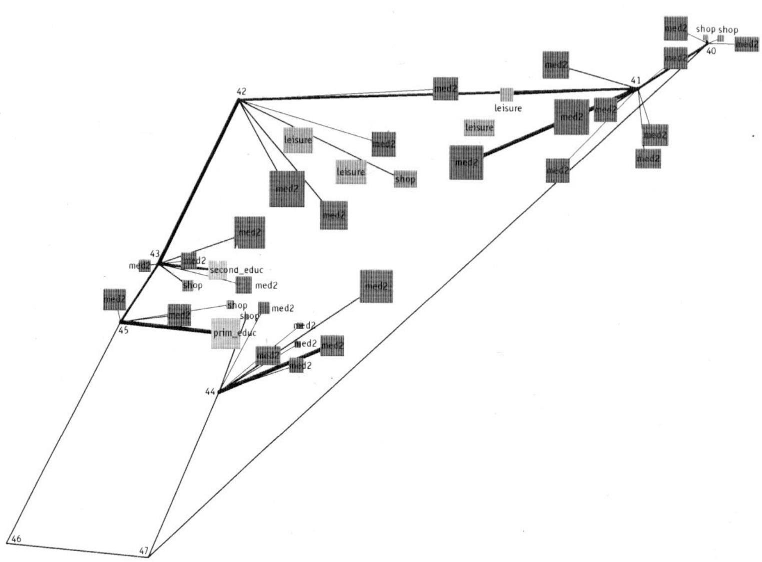

Source: Authors.
Note: To see this figure in color, refer to the appendix at the end of the book.

activities for the estimated population, but with no additional constraints on how these demands were met. The result, shown in figure 2.3, demonstrates that the optimizer found a solution that relied heavily on high-density housing because it provides accommodation in the most cost-effective manner. Similarly, the amount of open space provided was limited, and clusters of housing and activities can be seen gathered around each transport node (minimizing transport costs).

After this layout was discussed with the developers, it was clear that such extreme optimization would be undesirable in practice. For example, planning restrictions require a minimum total area of open space per capita, and the exclusive use of high-density housing would mean that the eco-town might not be a suitably attractive place to live. A second optimization therefore was run after adding corrective constraints on the maximum total area for high-density housing and adding green space and adjusting the minimum and maximum lot areas for mixed use and school facilities. The solution, illustrated in figure 2.4, shows a layout that is remarkably similar to the original master plan, although the housing density is still somewhat higher and some cells are unused.

The three scenarios are summarized in table 2.1. The results suggest that, when compared to the reference master plan case, an optimized layout could deliver up to a 60 percent reduction in total development costs and an approximately 80 percent reduction in energy consumption and carbon emissions. Even when constraints are

FIGURE 2.3
Cost-Optimized Eco-Town Layout (unconstrained)

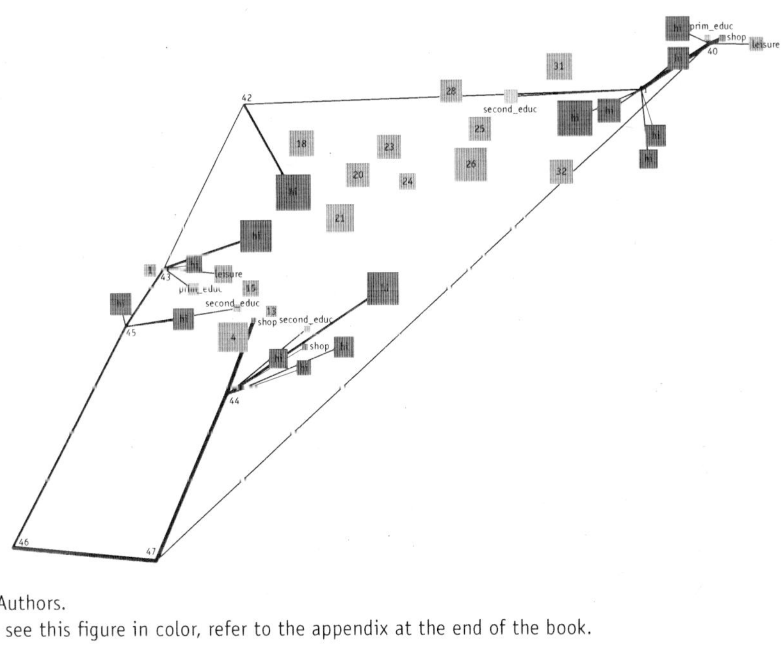

Source: Authors.
Note: To see this figure in color, refer to the appendix at the end of the book.

FIGURE 2.4

Cost-Optimized Eco-Town Layout (with additional planning constraints)

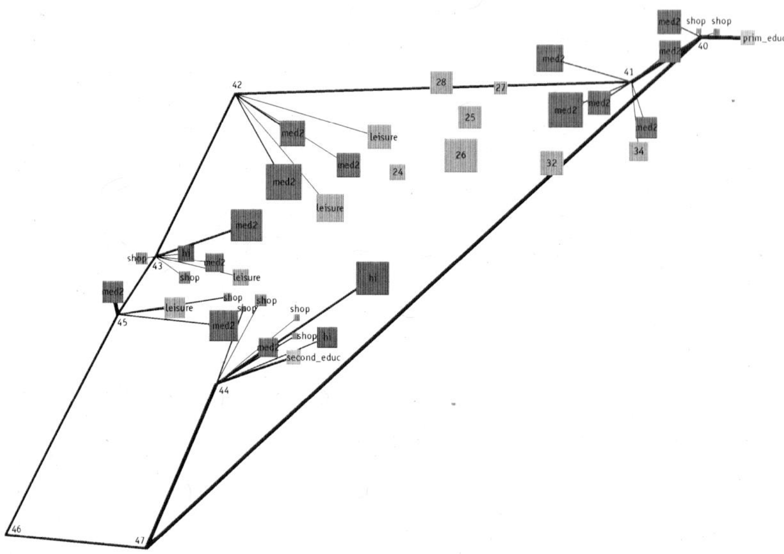

Source: Authors.
Note: To see this figure in color, refer to the appendix at the end of the book.

TABLE 2.1

Comparison of Layout Model Scenarios

	Master plan	*Unconstrained*	*Constrained*
High-density housing (% of total housing area)	0	100	15
Total housing provided (people)	8,297	6,576	6,760
Average daily travel (pass-km)	10,400	6,240	9,220
Network length (km)	19.9	14.7	17.4
Relative annual cost (Master plan = 100)	100	41	64
Annual energy consumption (GJ per capita)	86.2	19.3	52.3
Annual carbon emissions (tC per capita)	2.56	0.54	1.53

Source: Authors.
Note: GJ = gigajoules; tC = tonnes of carbon.

added to reflect the more realistic conditions of planning regulations and commercial viability, significant savings of approximately 40 percent are seen.

The layout model can also incorporate the provision of activities in "hinterlands" (for example, neighboring cities) and optimize for other goals (such as minimum carbon or energy consumption). Those results are not presented here.

Agent-Activity Model Results

A fully detailed agent-activity model is currently under development; however, this study was conducted using a less complicated version that requires very little computational overhead but must be tested to ensure that it generates realistic resource demands. (This case demonstrates the advantage of SynCity's modular approach: once the full version of the agent-activity model is complete, it will be able to directly replace the current simplified version with no additional effort because the inputs and outputs are identical.) Using the constrained layout calculated above, the agent-activity model was run to compare the calculated demands against those values estimated by the developers.

The model begins by assigning the population to houses and employers within the city layout. Time within the model is divided into 16 periods representing 4 times per day, 2 types of day (weekday, weekend), and 2 seasons (summer, winter). Each time period has different profiles for the types of activities that citizens might like to perform (for example, higher propensity for work during weekday daytime), and depending on their unique characteristics (for example, age, education, or income), a unique schedule of activities is chosen for each citizen. These activities are then performed and the associated demands for travel, heat, electricity, and other resources are calculated.

Table 2.2 summarizes the results of the model. The total demands for heat and electricity are very close to the reference values, although there are insufficient data available to ensure that the spatial and temporal distributions match. The daily trips generated by the agent-activity model are also similar to the reference case. However, there are insufficient data to verify the modal distribution; the agent-activity model

TABLE 2.2

Comparison of SynCity Agent-Activity Model Results to Eco-Town Reference

	SynCity	*Eco-town reference*
Annual heat demand (kWh per capita)	5,100	5,200
Annual electricity demand (kWh per capita)	2,500	2,080
Daily motorized trips (per household)	0.48	0.45

Source: Authors.

estimates that bus travel accounts for 94 percent of motorized trips, which likely underestimates the use of private cars. As a result, the demand for transport fuel calculated by the model (approximately 20 liters per person per year) was deemed to be unrealistic; therefore, the subsequent RTN analysis focuses only on the provision of heat and electricity.

Resource-Technology-Network Model Results

The RTN model is designed to select an optimal (in this case, lowest cost) energy-supply strategy for a pattern of resource demands. Four different supply strategies are available to meet a resource demand: (1) importing the resource from a hinterland, (2) receiving transfers of the resource from another location within the city, (3) creating the resource locally (for example, by converting other resources), and (4) retrieving stored resources.

Two simple cases are considered here. First, we assume that for a small development of 6,500 people, the business-as-usual case would involve importing all of the required resources from external hinterlands: in other words, importing gas and electricity from national grids. Small household-scale technologies (such as 20-kilowatt boilers) are then used to convert the gas into the required heat demands. The model is constrained to import resources to only one cell per resource, and a network configuration is then calculated to distribute these resources to their demand locations. Figure 2.5 shows the resulting layout for the gas and electricity networks.

In the second case, we note that the developers expressed interest in providing district heating at the site. We therefore restrict the model, forbidding it from importing electricity but allowing it to install a 50-megawatt combined-cycle gas turbine (CCGT) somewhere within the city, with an associated district heat network. Figure 2.6 shows the resulting networks for gas, electricity, and district heat. Clearly, district heat is provided only to those heat demands near the CCGT unit (using a heat exchanger to convert district heat to domestic grade heat), with gas being transported by pipeline to other locations and converted to heat in domestic boilers. Electricity is provided entirely by the CCGT and distributed to the points of demand.

In each plot, the gray boxes represent a location within the city, although not all cells have resource demands. Resource flows are shown by the arrows: bold vertical lines represent imports, and the width of the line is proportional to the resource flow. Labeled circles within a cell represent the presence of one or more conversion technologies.

Discussion and Conclusions

The examples above demonstrate the capabilities of the SynCity tool kit, a three-stage system for the modeling of urban energy systems. In the first layout step, it was shown how a mixed-integer linear programming model can be used to develop

FIGURE 2.5

Distribution Networks and Conversion Technologies for Gas and Electricity

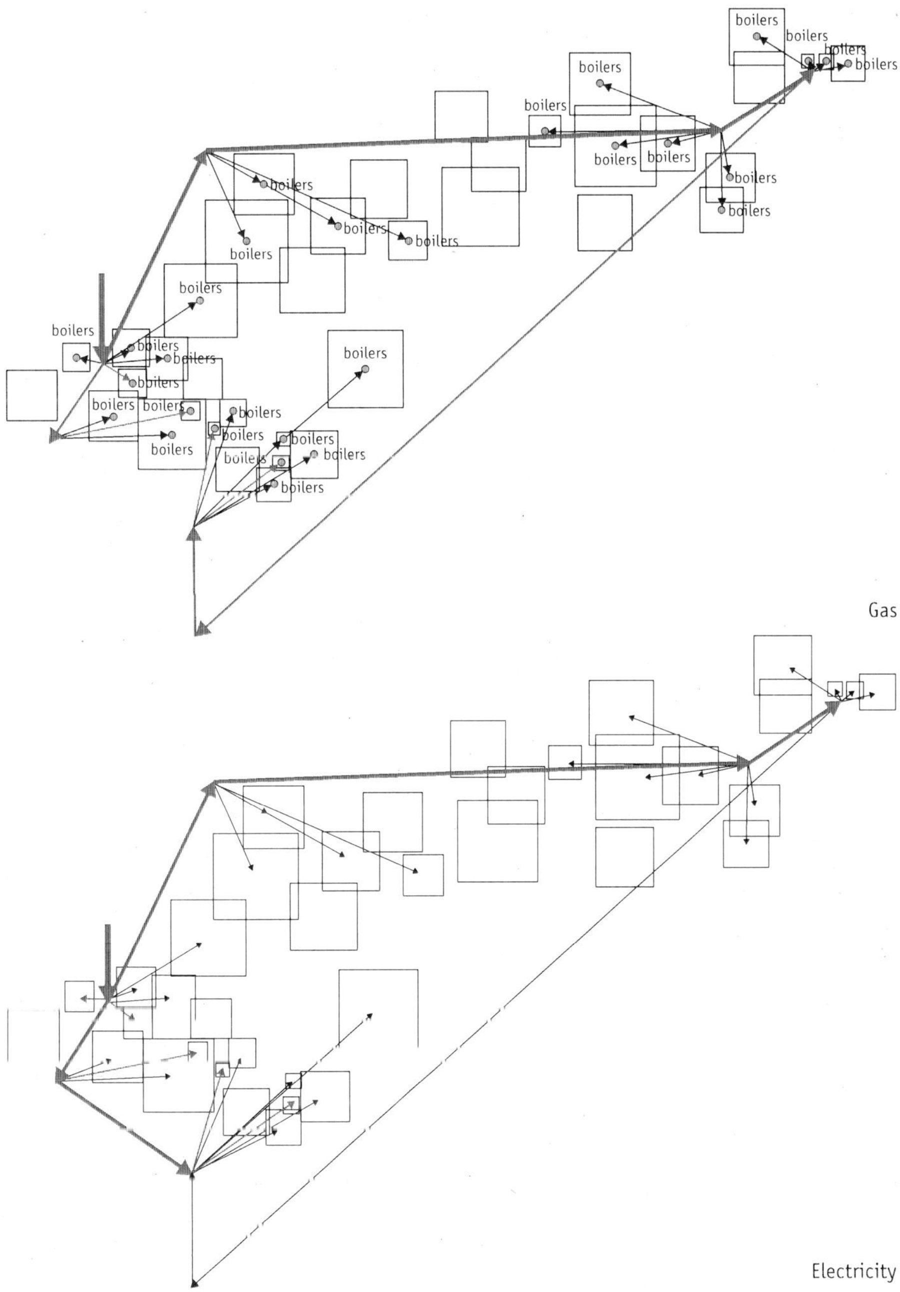

Source: Authors.

Note: Distribution networks and conversion technologies for gas and electricity in the import-only scenario. Boilers = household-scale gas boiler.

FIGURE 2.6

Distribution Networks and Conversion Technologies for Gas, Electricity, and District Heat

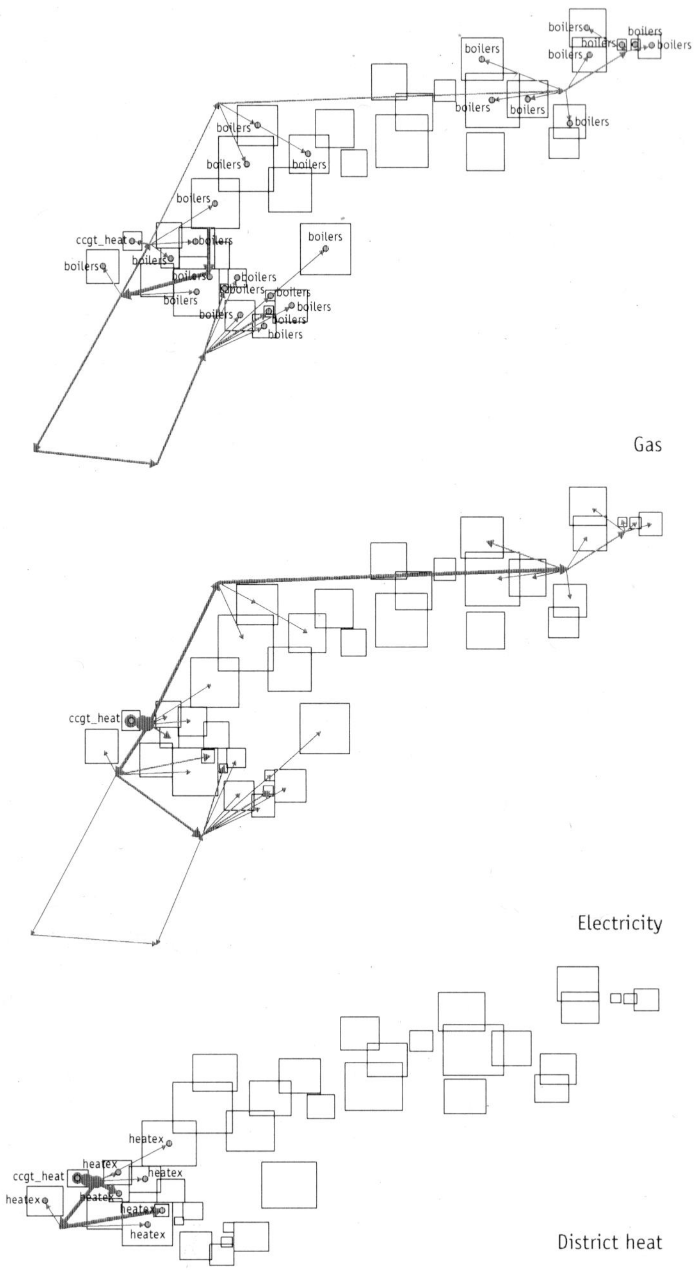

Source: Authors.

Note: Distribution networks and conversion technologies for gas and electricity in the import-only scenario. Heatex = heat exchanger, boilers = household-scale gas boiler, ccgt_heat = combined-cycle gas turbine with district heat take-off.

alternative master plans for a new development, with up to 80 percent reductions in cost and emissions against a business-as-usual scenario. With the addition of a few basic constraints, the design can be modified to reflect the requirements of a realistic developer while still delivering significant efficiency gains. Second, an agent-activity model was used to determine how individual agents might interact with this urban infrastructure, creating a spatially and temporally varied pattern of resource demand. While a more elaborate version of this model is currently being developed, the results calculated here are very similar to those assumed by the developer, indicating that the current model is a good placeholder implementation. Finally, a second optimization model was used to determine how these patterns of resource demand can be satisfied most cost-effectively using a combination of local conversion technologies and imported resources. The import-only and constrained-import examples demonstrate how the model framework can be easily adapted to assess alternative energy supply strategies.

Limitations of the Model

The case studied here represents a proposed eco-town development in the United Kingdom, but the SynCity system design is intended to be flexible so as to be applicable to other contexts without extensive customization. Indeed, in another case study, we were able to perform an initial assessment of a Chinese development in approximately two days, using data available from the project master plan. Nevertheless, the tool kit does have constraints and limitations, and to consider how these might affect the use of SynCity to developing country contexts, we briefly evaluated SynCity against a series of hypothesized use cases. The analysis led to the following conclusions:

- ***Themed-city assessments:*** The RTN model is built on a highly abstracted view of energy resources and technologies, and it can be restricted to consider only certain resources and conversion technologies. This facilitates the analysis of "themed cities." For example, in a developed country, this may mean the provision of biomass-fired district heating or a hydrogen city; in developing countries, technologies such as open fireplaces or kerosene burners could be simulated. Both examples require only minor changes to the model's input data (that is, descriptions of the cost and performance of these technologies) and, therefore, in this case, the difference between developed and developing country analyses is negligible.
- ***Impact assessments:*** The RTN model enforces a strict resource balance, which means that the outputs of fuel consumption must be explicitly accounted for. While waste heat is currently the primary waste output (for example, for assessing heat island impacts), other wastes (for example, tCO_{2e}, NOx, PM_{10}) can be included as additional resources that are generated by certain processes or agents' activities. Work is also currently under way to incorporate a full spectrum of global warming potential, resource depletion, and local air pollution

indicators in a more efficient manner. This is particularly relevant for developing countries where the use of lower grade energy supplies in urban areas creates significant health problems (Barnes, Krutilla, and Hyde 2005).

- ***Retrofit assessments:*** The current RTN model has a supply-side focus and assumes that any demand-side management measures have been undertaken in the agent-activity model. Therefore, another feature of the RTN model currently in development will allow an investment budget to be used for supply provision, demand management, or a mix of both measures. This will enable the model to compare retrofit programs, such as insulation drives, with the cost of providing additional energy supplies.
- ***Dynamic assessments:*** Although the models currently consider dynamic resource supplies and demands, the proposed layouts and technologies are all assumed to be static; that is, the model develops an optimized energy system at one point in time. However, for cities in developing countries that are undergoing rapid urbanization, there is a need to understand how infrastructure investments can be planned in stages. By using multiperiod optimization techniques, we hope to facilitate assessments of such phased energy infrastructure investments, as well as detailed agent behavior models to simulate the impact of policies over different time scales (for example, congestion pricing in transport on a daily scale, through to market transformation policies over a multiyear period).

These use cases demonstrate both the capabilities and limitations of the current SynCity system as it applies to developing countries. For some of these cases, such as multi-period investment decisions, new capabilities will need to be built into the tool kit. However for many problems, performing a developing country assessment is primarily a question of collecting the relevant technology and resource data. Again, SynCity is designed to have low data requirements and, as such, it is supported by a database of technologies and resources that largely remains static between countries. However, the cost of certain technologies and resources is likely to vary between countries, and this is an area where effort will be needed when transferring the system.

Recommendations

The goal of the BP Urban Energy Systems project is to identify ways of achieving a 50 percent improvement in the energy intensity of cities. The SynCity tool kit described here facilitates that goal by providing a common modeling framework for a full range of urban energy modeling problems, such as master planning, citizen behavior, and resource provision. At present, the tool kit should be used as a starting point for discussions about how to achieve this goal. The optimization techniques used in the layout and RTN models make it possible to identify and quantify significant energy savings, but these solutions must be tempered with the requirements of developers and citizens. Therefore, we recommend that the tool kit be used first to establish a baseline for any business-as-usual plans and then to quantify and compare the results of more optimized solutions. As demonstrated by the layout model, constraints can be added iteratively in discussion with developers and policy makers to identify effective compromises.

A policy-relevant example of how this process might work can be found in efforts to assess the energy implications of planning constraints. For example, a certain amount of urban green space is desirable to provide both amenity and ecosystem services; however, large amounts of green space decrease the density of urban form, thereby increasing transportation costs and reducing the feasibility of systems such as district heating. In this context, SynCity could be used to examine the energy impacts of different planning constraints on the provision of green space. The layout model would determine how these green space constraints can be satisfied while still minimizing transport energy demand; and the RTN model would subsequently develop an optimized energy strategy for this spatial configuration. The results of multiple scenarios could be compared in discussion with policy makers to decide which planning policy is most appropriate.

Ultimately, we envisage SynCity being used in three different ways. First, a lightweight version of the tool kit could be used to rapidly assess the major trade-offs within urban energy systems (for example, density effects and centralized versus decentralized energy systems). This version could be developed as a Web tool and supported by illustrative case studies, enabling policy makers without expertise in energy systems to get an understanding of the key issues. The second application would be a more detailed design tool. Here, expert policy makers or designers could use the optimization models to explore the limits of low-energy urban design. Finally, the underlying database of urban energy resources and technologies offers a useful reference tool. These three components—the light version of the modeling tool, the full version of the modeling tool, and the underlying database—therefore comprise the potential deliverables of the SynCity system.

Policy makers, engineers, and other stakeholders face a wide variety of urban energy challenges. Although still in a prototype stage, the SynCity tool kit offers a powerful platform for the integrated modeling of urban energy systems. While other systems will still be required for microassessments, the capabilities presented here show that SynCity uniquely allows users to evaluate holistic urban energy strategies from the early master plan stage through to assessing the impacts of a specific energy supply strategy. Development of the platform continues, with a view to providing a stable tool kit to users in both the developed and developing world.

Note

1. Because of commercial confidentiality concerns, it was difficult to get accurate cost data for each building and infrastructure type; the results, therefore, compare costs as normalized against this modeled baseline.

References

Barnes, D. F., K. Krutilla, and W. F. Hyde. 2005. *The Urban Household Energy Transition: Social and Environmental Impacts in the Developing World*. Washington, DC: Resources for the Future.

Batty, M. 2007. "Urban Modeling." In *International Encyclopedia of Human Geography*, ed. R. Kitchin and N. Thrift. London: Elsevier.

BioRegional and CABE (BioRegional Development Group and the Commission for Architecture and the Built Environment). 2008. "What Makes an Eco-town? A Report from BioRegional and CABE Inspired by the Eco-towns Challenge Panel." Commission for Architecture and the Built Environment, London.

Brownsword, R. A., P. D. Fleming, J. C. Powell, and N. Pearsall. 2005. "Sustainable Cities—Modeling Urban Energy Supply And Demand." *Applied Energy* 82 (2): 167–80.

CUSPA (Center for Urban Simulation and Policy Analysis). 2009. *UrbanSim*. University of Washington: CUSPA. http://www.urbansim.org/, accessed April 14, 2009.

Dawson, R. J., J. W. Hall, S. L. Barr, et al. 2009. "A Blueprint for the Integrated Assessment of Climate Change in Cities." Tyndall Centre Working Paper 29, United Kingdom. http://tyndall.ac.uk/sites/default/files/wp129.pdf, accessed April 14, 2009.

DCLG (Department of Communities and Local Government). 2008. "Sustainability Appraisal and Habitats Regulations Assessment of the Draft Planning Policy Statement and the Eco-towns Programme." DCLG, London.

Feng, C- M., and J- J. Lin. 1999. "Using a Genetic Algorithm to Generate Alternative Sketch Maps for Urban Planning." *Computers, Environment and Urban Systems* 23 (2): 91–108.

Girardin, L., M. Dubuis, N. Darbellay, et al. 2008. "Energis: A Geographical Information Based System for the Evaluation of Integrated Energy Conversion Systems in Urban Areas." ECOS 2008: 21st International Conference on Efficiency, Cost, Optimization, Simulation and Environmental Impact of Energy Systems, Kraków, Poland, June 24–27.

IEA (International Energy Agency). 2008. *World Energy Outlook 2008*. Paris: International Energy Agency/Organisation for Economic Co-operation and Development.

ISCCC (International Scientific Congress on Climate Change). 2009. "Climate Change—Global Risks, Challenges and Decisions: Key Messages from the Congress." Copenhagen, March 10–12. http://climatecongress.ku.dk/newsroom/congress_key_messages/, accessed April 14, 2009.

Mori, Y., Y. Kikegawa, and H. Uchida. 2007. "A Model for Detailed Evaluation of Fossil-Energy Saving by Utilizing Unused But Possible Energy-Sources on a City Scale." *Applied Energy* 84 (9): 921–35.

Noy, N. F., and D. L. McGuinness. 2002. "Ontology Development 101: A Guide to Creating Your First Ontology." http://protege.stanford.edu/publications/ontology_development/ontology101.pdf, accessed October 14, 2009.

Robinson, D., N. Campbell, W. Gaiser, et al. 2007. "SUNtool—A New Modeling Paradigm for Simulating and Optimising Urban Sustainability." *Solar Energy* 81 (9): 1196–211.

Shah, N., D. Fisk, A. Davies, et al. 2006. "BP–Imperial College Urban Energy Systems Project: First Annual Report." http://www3.imperial.ac.uk/pls/portallive/docs/1/23275696.PDF, accessed October 14, 2009.

Tweed, C., and P. Jones. 2000. "The Role of Models in Arguments About Urban Sustainability." *Environmental Impact Assessment Review* 20 (3): 277–87.

UN (United Nations). 2008. "World Urbanization Prospects: The 2007 Revision Population Database." New York: United Nations Population Division. http://esa.un.org/unup/, accessed August 7, 2008.

Urban, T. L., W-C. Chiang, and R. A. Russell. 2000. "The Integrated Machine Allocation and Layout Problem." *International Journal of Production Research* 38 (13): 2911.

CHAPTER 3

City-Scale Integrated Assessment of Climate Impacts, Adaptation, and Mitigation

Jim W. Hall, Richard J. Dawson, Stuart L. Barr, Michael Batty, Abigail L. Bristow, Sebastian Carney, Athanasios Dagoumas, Alistair C. Ford, Colin Harpham, Miles R. Tight, Claire L. Walsh, Helen Watters, and Alberto M. Zanni

Worldwide, cities are faced with the challenge of designing and implementing the transition to a state in which their greenhouse gas emissions are drastically reduced and they are well adapted to the effects of climate change. There have now been several studies of the synergies and conflicts in the objectives of mitigation and adaptation. These interactions are most vivid in urban areas, where they play out through land use, infrastructure systems, and the built environment. Urban decision makers need to understand the implications of these interactions and the potential influences of future global climate changes. With these decision makers in

---

The research described in this paper was funded by the Tyndall Centre for Climate Change Research, United Kingdom. Jim W. Hall is Professor of Earth Systems Engineering, School of Civil Engineering and Geosciences, Newcastle University, United Kingdom. Richard J. Dawson is EPSRC Research Fellow, School of Civil Engineering and Geosciences, Newcastle University, United Kingdom. Stuart L. Barr is Senior Lecturer in Geographic Information Science, School of Civil Engineering and Geosciences, Newcastle University, United Kingdom. Michael Batty is Professor of Planning, Centre for Advanced Spatial Analysis, University College London, United Kingdom. Abigail L. Bristow is Professor of Transport Studies, Department of Civil and Built Environment, Loughborough University, United Kingdom. Sebastian Carney is Researcher, School of Environment and Development University of Manchester, United Kingdom. Athanasios Dagoumas is Senior Energy Analyst, Hellenic Transmission System Operator S.A., Greece. Alistair C. Ford is Researcher, School of Civil Engineering and Geosciences, Newcastle University, United Kingdom. Colin Harpham is Senoir Research Associate, Climatic Research Unit, University of East Anglia, United Kingdom. Miles R. Tight is Senior Lecturer in Transport Planning and Engineering, Institute for Transport Studies, University of Leeds, United Kingdom. Claire L. Walsh is Researcher,

(*continued on next page*)

mind, the Tyndall Centre for Climate Change Research, since 2006, has been developing an Urban Integrated Assessment Facility (UIAF), which seeks to simulate socioeconomic change, climate impacts, and greenhouse gas emissions over the course of the twenty-first century. The research focuses on London, a city that has taken a lead role in the United Kingdom and globally with respect to climate protection. The objective of this chapter is to provide an overview of the structure of the Tyndall Centre UIAF and a brief description of the various component modules and sample results. In this context, the application of the model to London and the way in which it is being used to analyze policy options in the city are discussed.

Introduction

Responding to the threats of climate change by adopting strategies to mitigate greenhouse gas (GHG) emissions and to adapt to the effects of climate change is placing new and complex demands upon urban decision makers, who simultaneously function against a background of rapid socioeconomic change. Determining targets for mitigation of GHG emissions is now urgent and will require a major reconfiguration of urban energy and transport systems as well as the built environment. Developing these adaptations requires integrated thinking that encompasses a whole range of urban functions. Further, the climate changes that require adaptation, which manifest now as intermittent extreme climate shocks of floods, droughts, windstorms, or heat waves, can be expected to amplify in the coming decades. The timescales of change and the variability associated with the change signals are quite out of step with most political decision-making processes.

Mitigating GHG emissions and adapting to climate change in urban areas involves complex interactions of citizens, government and nongovernmental organizations, and businesses. This complexity can inhibit the development of integrated strategies involving demand management, land use planning, construction of new civil infrastructure, and so on, whose combined effect is potentially more beneficial than the achievements of any single agency or organization acting unilaterally.

The understanding of the synergies and conflicts in the objectives of mitigation and adaptation has been increasing (McEvoy, Lindley, and Handley 2006). These interactions are most evident in urban areas, where they play out through land use, infrastructure systems, and the built environment. Without sensible planning, climate change can induce energy-intensive adaptations, such as air conditioning or desalination, driving higher GHG emissions. Urban decision makers need to understand the implications of these interactions and the potential impacts of future global changes in climate. If they can test alternative policies through simulation and assess their multiple attributes, they are more likely to avoid the mistakes of the past.

(continued from previous page)

School of Civil Engineering and Geosciences, Newcastle University, United Kingdom. Helen Watters is Researcher, Institute for Transport Studies, University of Leeds, United Kingdom. Alberto M. Zanni is Researcher, Department of Civil and Built Environment, Loughborough University, United Kingdom.

While the objectives of mitigation and adaptation are clear and well aligned with the broader aims of sustainable development, the process of designing transitions to sustainability in urban areas is much more complex. In practice, these objectives will be achieved by a myriad of local actions set within a broad policy framework. On the other hand, certain large-scale infrastructure and planning decisions are essential elements within the portfolio of measures that need to be established as part of a transition strategy. Cities, therefore, are regarded as complex adaptive systems over which urban development decision makers have only partial control. The present urban configuration can only be understood in the context of past development, by which it arrived in its current state. In other words, the state of cities is "path dependent." By the same token, future development options are modulated by existing development on the ground and development trajectories.

The authors contend that the processes of influence and interaction within urban areas are so complex that they increasingly defy the ability of individual decision makers to assimilate all of the relevant information and reach strategic decisions. The well-articulated aspiration to produce an integrated response to the challenges facing cities may in practice be overwhelmed by the complexity of factors that must be considered in these decisions. Therefore, methods and tools that can help to facilitate and inform integrated assessment are needed. Quantified integrated assessment described in this chapter is one such approach.

Integrated assessment has been applied to a wide variety of different systems and for a range of spatial and temporal scales. The application here is for cities, their long-term processes of change, and how climate-related drivers interplay with other drivers (for example, demographic and economic processes of change) over timescales of up to 100 years into the future. This extended period coincides with the typical timeframe for assessment of climate change policy. An extended timeframe is also motivated by the long life of infrastructure systems and the extended legacy of planning decisions. It is these major planning and design decisions that this assessment seeks to inform so as to avoid decisions whose consequences are materially regrettable or that foreclose the opportunity for alternative actions in the future. However, given the major uncertainties over such an extended timescale, careful attention has to be paid to analysis and representation of uncertainties. These are mostly dealt with by using sets of future scenarios. Probabilistic information on climate uncertainties has recently become available in the United Kingdom (UK) thanks to the latest climate change scenarios (UKCP09, the fifth generation of climate information for the UK).

If processes of change with respect to extended timescales are to be understood, then it is usually also necessary to analyze them on extended spatial scales. Therefore, this analysis is conducted on the scale of whole cities, the scale at which patterns of spatial interaction are most vivid. However, framing the city in this way would cause inevitable boundary problems that will be evident depending on which economic and transport interactions between the metropolis and the surrounding region and nation the user seeks to represent and assess.

Similarly, to analyze water resources and flooding, the whole of the surrounding river basin, together with interbasin transfers where they exist, needs to be examined.

Alongside these factors, there are other aspects of urban climate that require a nested approach to downscaling from the global climate. The study boundaries that are set are therefore multiple and noncoinciding, though they all have a certain rationale.

The approach described in this chapter is a work in progress. A first phase of development, which is now approaching completion, has put in place all the main components that are required for an integrated assessment of mitigation and adaptation in urban areas; an overview of this is provided in the next section, "The Integrated Assessment Framework on Climate Change and Cities: Approach Adopted." The work is now being used in practice to help inform decision making in London, although scope for refinement remains. In this context, climate-related complexities that call for an integrated assessment approach in London and the Thames Gateway are highlighted in the third section. The fourth section describes, with the help of some indicative demonstration datasets and outputs, how the integrated assessment is being constructed, while the fifth and sixth sections indicate the potential of the integrated assessment framework. The assessment approach adopted shows that it is both possible and informative to look at cities in an integrated and quantified manner over a timescale of decades. Thus, as the seventh section concludes, this framework for analysis is proving to be of value in stimulating more integrated thinking about cities in the context of climate change and informing complex decision-making problems.

The Integrated Assessment Framework on Climate Change and Cities: Approach Adopted

In 2006, the Tyndall Centre for Climate Change Research launched a research program that is developing a quantified integrated assessment model for analyzing the impacts of climate change on cities and their contribution to global climate change in terms of GHG emissions. The overall approach uses quantified scenario analysis, which combines a number of modules that represent relevant processes of long-term change in urban areas.

Cities provide a scale at which strategies for mitigation and adaptation to climate change can be usefully designed and assessed within a quantified assessment framework. Increasingly, this is also the scale at which individual civil servants in city administrations are being given responsibility for climate protection. Yet urban climate mitigation and adaptation policy and behavior can hardly be divorced from their global context. Therefore, a nested approach, in which a wide range of global climate, economic, and demographic scenarios are taken as the boundary conditions for the integrated assessment, has been adopted.

The overall structure of the integrated assessment discussed here is shown in figure 3.1. The approach begins with an initial analysis of socioeconomic and climate scenarios that form the boundary conditions for the analysis. A process of downscaling generates climate scenarios at the city scale, as well as economic and demographic scenarios for the urban area. This provides the boundary conditions for the city-scale analysis in this case study for the city of London. A spatial interaction

FIGURE 3.1
Overview of the Integrated Assessment Methodology for Greenhouse Gas Emissions and Climate Impacts Analysis at a City Scale

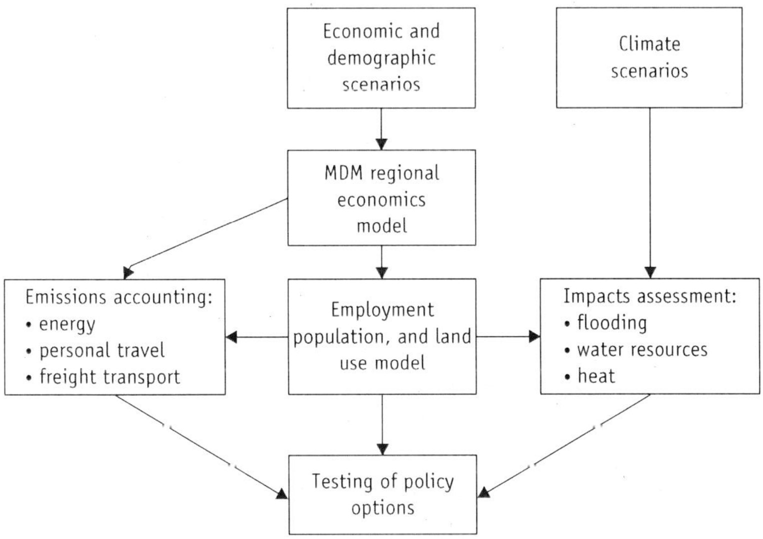

Source: Authors.
Note: MDM = Multisectoral Dynamic Model.

module provides high-resolution spatial scenarios of population and land use that form the basis for analysis of GHG emissions and climate impacts. The modules for emissions accounting and climate impacts analysis are depicted on the left and right sides of figure 3.1, respectively. These provide projections of emissions and climate impacts under a wide range of scenarios corresponding to climate, socioeconomic, and technological changes. The Urban Integrated Assessment Facility (UIAF) provides the flexibility to test a very wide range of mitigation and adaptation policies, including land use planning, modifications to the transport systems, changing energy technologies, and measures to reduce climate risks. These portfolios of policies are brought together in a portal for end users, depicted in the bottom panel of the figure.

The framework set out in figure 3.1 is intended to be generic for climate impacts and GHG emissions analysis at a city scale. The component models presented in subsequent sections are also generic in that they can, and in most cases have been, demonstrated with respect to other case study sites. However, the application of the approach to other locations is conditional on the availability of the necessary data. Key datasets required for the integrated analysis include land use, census, and economic information, along with information on transport and other infrastructure and travel behavior. While the datasets are commonly available in OECD (Organisation for Economic Cooperation and Development) countries, obtaining the necessary datasets elsewhere will be more arduous.

The Tyndall Centre Cities Programme, in operation since 2006, was initially begun with the assembly of existing modules that could form part of the integrated assessment, including economics, climate downscaling, and emissions accounting modules. A framework for transfer of state variables between modules was subsequently established, before development began of new modules for land use, transport, and climate impacts analysis. In recent months, these have been brought together into the integrated assessment framework described here.

Greater London and the Thames Gateway: Vulnerable to a Changing Climate

The capital of the United Kingdom, London has been settled for around two millennia. With a wide cultural, social, economic, environmental, and built heritage, London is one of the most diverse cities in the world, with 29 percent of the population from ethnic minorities and speaking almost 300 languages (ONS 2003). The current population is approximately 7.2 million and is expected to grow to more than 8.1 million by 2016 (GLA 2004). The London Plan (GLA 2004) is a strategic plan developed by the Greater London Authority (GLA) setting out an integrated social, economic and environmental framework for the future development of London over the next 15 to 20 years. The plan provides the citywide context within which individual boroughs (local administrative authorities, of which there are 33 in London) must set their local planning policies. The plan's general objectives are to

- accommodate growth within current boundaries without encroaching on open spaces.
- make London a "better" city to live in.
- strengthen and diversify economic growth.
- increase social inclusion and reduce deprivation.
- improve accessibility through use of public transport, cycling, and walking (that is, reduce the use of cars, though airport, port, and rail infrastructure are likely to be increased).
- make London a more attractive, well-designed, green city through improved waste management, reuse of "brownfield" sites, increased self-sufficiency, and improved air quality.

The London Plan sets out areas targeted for development, with an emphasis on redeveloping previously developed land and certain areas that are targeted for regeneration (figure 3.2). A series of "metropolitan centers" have been identified as focal points of economic activity. Economically depressed areas have been identified as "regeneration areas," which are concentrated in east London and include the 2012 Olympics site. The metropolitan area of London extends beyond the boundaries of the GLA. Of particular significance is the zone of development east of London along the estuary of the river Thames, known as the Thames Gateway (figure 3.2), which includes many previously industrialized areas that are now underdeveloped but benefit from being quite close to central London and from having good transport links.

FIGURE 3.2
Zones of Development in London and the Thames Gateway

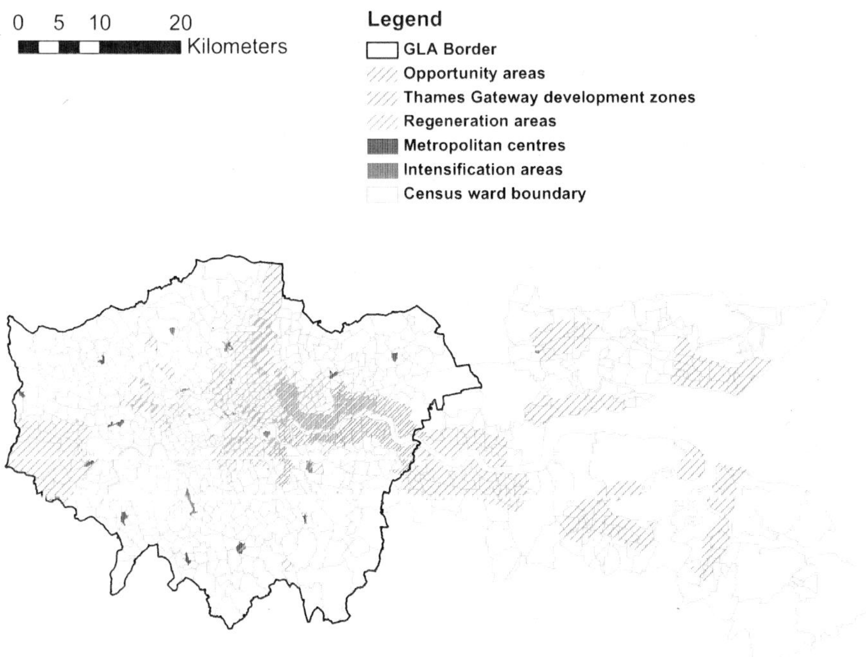

Source: Zones of development, opportunity areas, urban green space, and previously developed land within the GLA and Thames Gateway are from the Greater London Authority, http://www.london.gov.uk/thelondonplan/.
Note: To see this figure in color, refer to the appendix at the end of the book.

The southeast of the UK, where London is located, is particularly vulnerable to changing climate, being relatively water scarce and vulnerable to both sea level rise and storm surges. It is the most water-scarce region in the UK, having lower than average rainfall and a very large demand for water (Environment Agency 2007). Isostatic subsidence[1] means that southern Britain has experienced and will continue to experience faster relative sea level rise than in the north, and the southern North Sea has always been subjected to storm surges. While the evidence for future changes in surge processes is ambiguous, when superimposed on increased mean sea level, the risk of surge flooding will increase (Evans et al. 2004). Also, because of the concentration of population and transport activity in the area, the southeast region of the country is a focal point of GHG emissions. London in particular suffers from urban heat and associated air quality problems because of its geographical location in the warmer part of the UK and very widespread urbanization (London Climate Change Partnership 2002). In this context, the London Underground subway system requires adaptation measures, specifically air conditioning during summer, since it was not designed for these increasingly hot conditions.

Responding to the challenges of a changing climate requires a portfolio of measures that may involve the reversal of entrenched patterns of demand and development. Also, the need to adapt to climate change may conflict with the demands of mitigation (Klein, Schipper, and Dessai 2003). For example, higher urban temperatures may increase energy demand for air conditioning, which in turn would increase heat emissions into urban areas, thus exacerbating the problem. Action to tackle climate change needs to be set in the broader context of sustainability, including issues of resource use, human wellbeing, and biodiversity (Najam et al. 2003).

It is clear that both mitigation and adaptation will involve developing portfolios of measures in a strategic way with respect to issues administered by different agencies. For example, flooding in urban areas is subject to rather complex administrative arrangements in the UK (Pitt 2008), but responsibility for tidal flooding and flooding from the river Thames and its tributaries rests with the Environment Agency. Similarly, water resources are planned and managed by negotiation between privatized water utilities, the water regulator (OFWAT), and the Environment Agency. Many relevant instruments, such as duties on fuel, utility pricing, and the generation mix for national grid electricity generation, are in the hands of central government or privatized utilities and industry regulators, while local planning decisions are administered by the London boroughs. However, despite the complex administrative arrangements, the GLA's strategic planning responsibility in the context of land use planning, local regulations, and roads and public transport (via Transport for London[2]) provides it with opportunities to reduce emissions and climate vulnerability.

The complex institutional landscape outlined above makes it necessary to assemble various coalitions of stakeholders in support of mitigation and adaptation initiatives. Effective action can seldom be mobilized by one organization acting independently. However, collective action depends on developing collective understanding from a variety of perspectives, and computer modeling and decision support platforms can help to develop that collective understanding. The integrated assessment under development in the Tyndall Centre for Climate Change Research is intended to provide such a platform, by estimating current and future climate risks and emissions and providing the capacity to analyze alternative management options.

Elements of the Integrated Assessment

In the following sections, we describe the main elements of the integrated assessment for London that was introduced above.

Regional Economic Modeling

A regional economic model was used to provide the quantified economic scenarios that are the starting point for analysis of vulnerability and GHG emissions. The Multisectoral Dynamic Model (MDM) (Barker and Peterson 1987; Junankar, Lofsnaes, and Summerton 2007) has been adopted for the integrated assessment with respect to London. MDM is a coupled macroeconomic model because it models the whole economy, but is multisectoral, predicting output from and employment in 42 different industrial sectors. The version of the model used in this assessment contains each of the regions in the UK (including Greater London) as well as the rest of the world. It uses regional multisectoral economic data, which is available for the UK from the Office of National Statistics.

MDM models growth and dynamics over the medium and long term, so it is well suited to the task of providing internally consistent scenarios for the purposes of integrated assessment. The model is dynamic, providing intermediate results at time-steps over the simulation period. It takes as its inputs baseline projections of long-term growth and population, as well as past observations of relationships between different industrial sectors.

MDM has been used since the 1980s to provide economic forecasts for the UK economy, thus benefiting from a process of ongoing testing and validation. Providing projections for the timescale used in this analysis is obviously ambitious, so the outputs of the type shown in figure 3.3 should be understood as possible scenarios. These

FIGURE 3.3

Employment Projections for London, Comparing Outputs from the MDM Model (labeled "Tyndall") and Figures Used by the GLA

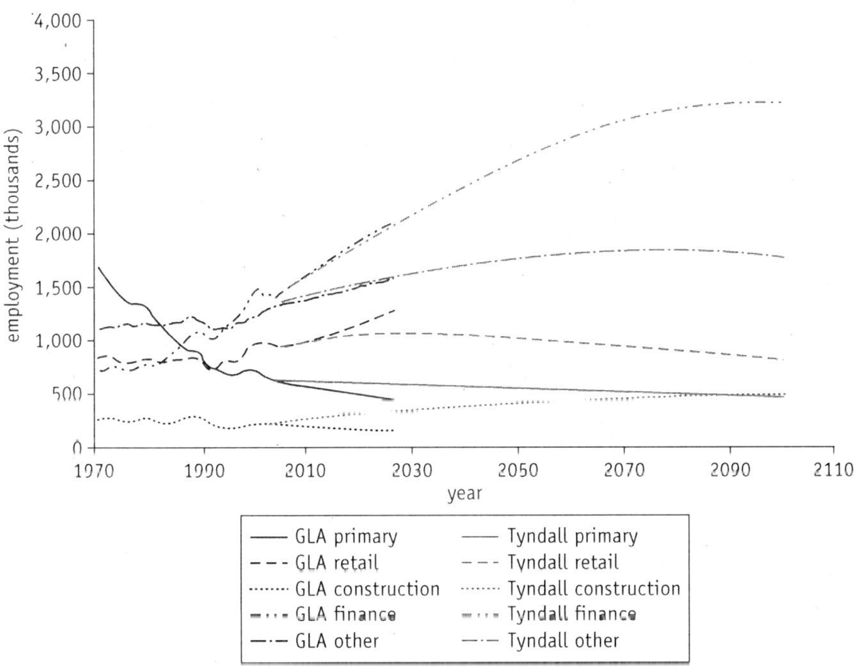

Sources: Trend-based employment forecasts for London by borough, Greater London Authority, http://www.london.gov.uk/mayor/economic_unit/docs/ep-technical-paper-1.pdf; current and historic population demographics, Office for National Statistics, http://www.statistics.gov.uk/default.asp; and future population demographics, exchange rates, interest rates, GDP growth, energy demand, national taxation, and government expenditure are from *Special Report on Emissions Scenarios* (SRES), http://www.ipcc.ch/.
Note: Industrial activity has been aggregated into five sectors.

scenarios, however, have the attraction of providing internally consistent projections, which account for intersectoral interactions and temporal dynamics, and provide inputs to the land use model and the emissions accounting models described below. In figure 3.3, economic activity has been aggregated into five sectors (from the 42 sectors included in MDM). For comparison, the Tyndall Centre scenarios generated in MDM are illustrated alongside a separate set of scenarios generated by the GLA up to the year 2025. The directions of both of these sets of scenarios are consistent, with the exception of the construction sector, which the GLA projects to shrink over the coming years, notwithstanding the volume of construction work associated with the 2012 Olympics and the Crossrail east-west underground mass transit project.

Land Use Change Modules

Future vulnerability to climate change and demand for energy services that emit GHGs are closely linked with land use patterns in urban areas. Moreover, spatial analysis of GHG emissions can help to target mitigation action and may also be a requirement for analysis of air quality and anthropogenic heating contributions to the urban heat island. Therefore, central to the UIAF are modules for the development of land use change scenarios over the coming decades. The starting point for this analysis is the current configuration of domestic and commercial land use, along with census data. The evolution of land use from the current configuration into the future is analyzed on the basis of a number of drivers and constraints. The multisectoral employment scenarios discussed above are disaggregated to provide spatial scenarios of future employment. Access to employment provides the main driver for development of new residential property. Generalized travel cost has been analyzed for five modes (road, bus, rail, light rail, and underground metro) to and from each of the 801 wards under analysis in the Greater London and Thames Gateway area. The generalized cost incorporates the cost for travel time (including walking to public transport nodes and waiting), along with ticket and fuel costs as appropriate. Note the influence of the congestion charging zone in central London on travel cost.

This formula is based upon analysis of the transport network and published data on travel times. Figure 3.4 shows an example of generalized travel costs by car (in minutes) from Heathrow ward. These generalized costs have been computed for each combination of origins and destinations in the spatial interaction model. The generalized travel costs have been modified to represent planned transport infrastructure investment (including Crossrail) and also to incorporate various scenarios of transport infrastructure development looking further into the future.

These travel costs are incorporated in a spatial interaction model, alongside information on existing land use and land available for new residential development, in order to develop future distribution of employed population and their dependents. Figure 3.5 illustrates the projected change (in 2100) in residential population at a ward scale for a high-growth scenario, assuming *unconstrained* development. In practice, patterns of commercial and residential land use will be constrained by planning policies, which, for example, may prevent development on recreational land. On the

FIGURE 3.4
Example of Generalized Travel Costs by Car (in minutes) from Heathrow Ward

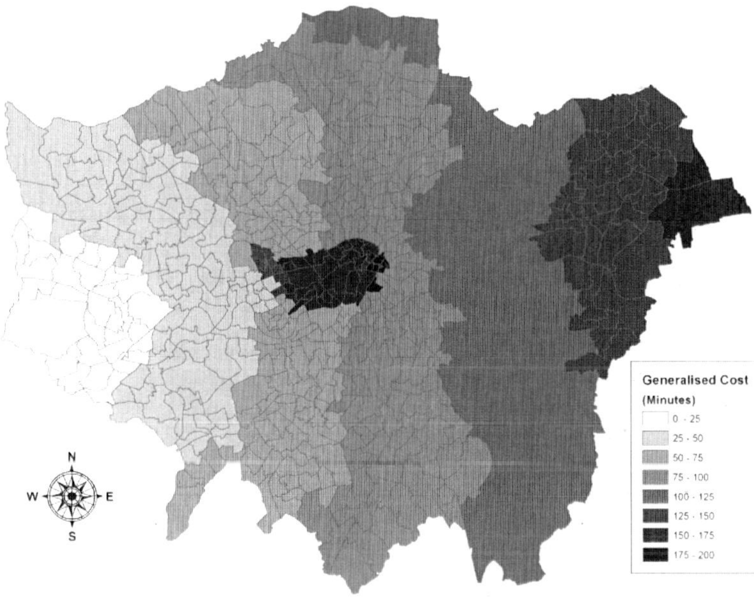

Sources: Road network. OS Mastermap ITN Data, http://www.ordnancesurvey.co.uk/oswebsite/products/osmastermap/layers/itn/; the London Travel Report, Transport for London, http://www.tfl.gov.uk/corporate/about-tfl/publications/1482.aspx.
Note: Note the influence of the congestion charging zone in central London on travel cost. To see this figure in color, refer to the appendix at the end of the book.

other hand, planning policy may actively seek to promote development in particular areas, for example "brownfield" sites of previously developed industrial land. These additional constraints and attractors can be applied to the spatial interaction model, with varying degrees of policy effectiveness. Figure 3.6 illustrates a population change scenario in 2100 subject to existing constraints upon land use and attractors for specified regeneration zones.

The spatial interaction model described here has been developed at the scale of wards, across the whole of London and the Thames Gateway. This keeps the analysis of interactions to a manageable 801 zones (633 for London and 168 for the Thames Gateway). However, for climate impacts analysis, higher resolution scenarios are required. A module that disaggregates land use development in a spatially explicit manner at the ward scale onto a 100 meter grid has therefore been developed. This module combines a series of weighting and constraint functions that project land use change alongside existing development, the pattern of which is available from land use classification maps. The weighting functions seek to locate residential properties near existing residential development, transport links, and schools, while applying

FIGURE 3.5

Projected Population Change in 2100 at a Ward Scale (high economic growth, unconstrained development)

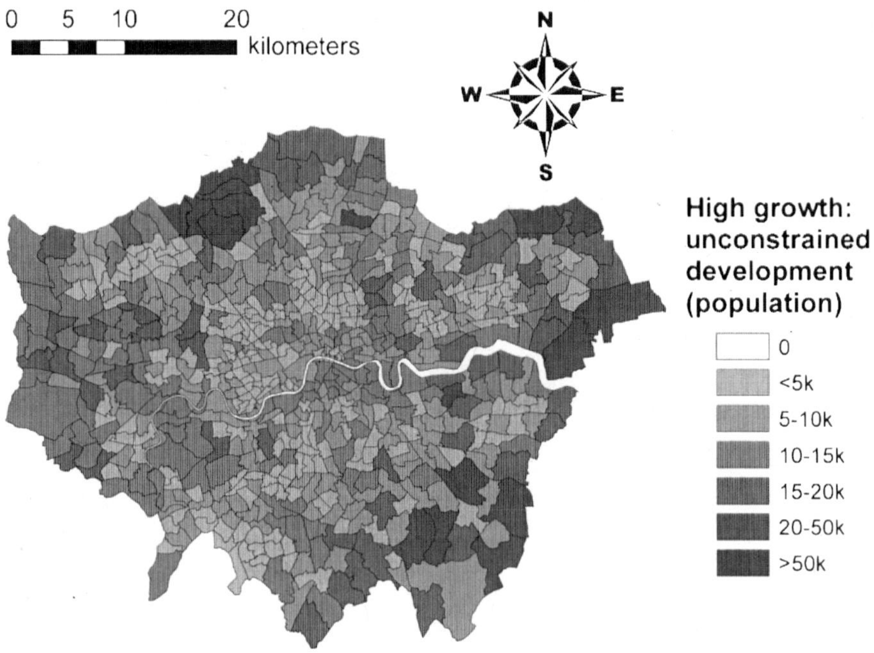

Sources: Trend-based employment forecasts for London by borough, Greater London Authority, http://www.london.gov.uk/mayor/economic_unit/docs/ep-technical-paper-1.pdf; current and historic population demographics, Office for National Statistics, http://www.statistics.gov.uk/default.asp; wards: census area statistics, U.K. Borders, http://edina.ac.uk/ukborders/; census data, U.K. Census Service, www.census.ac.uk; current land development, MasterMap, http://www.ordnancesurvey.co.uk/oswebsite/products/osmastermap/; index of deprivation, communities and local government, http://www.communities.gov.uk/communities/neighbourhoodrenewal/deprivation/deprivation07/; and property type and location, National Property Database, The Environment Agency, http://www.environment-agency.gov.uk.
Note: To see this figure in color, refer to the appendix at the end of the book.

local planning constraints. Figure 3.7 shows the results of the disaggregation algorithm for one ward in east London (South Hornchurch) under two different scenarios of population change. In both cases, development in brownfield sites is encouraged. It is clear from the figure how the overall growth scenario modifies the change in land cover at a ward scale. Figure 3.7 also illustrates how the ward-scale disaggregator locates new development near existing development. However, it is not possible to see from the figure how other attractors, such as proximity to transport nodes, have modified the pattern of local development.

Climate Impacts and Adaptation Analysis

The climate impacts analysis currently focuses on the three most important potential impacts of climate change in London: flooding, water scarcity, and heat waves.

FIGURE 3.6
Projected Population Change in 2100 at a Ward Scale (high economic growth, constrained development)

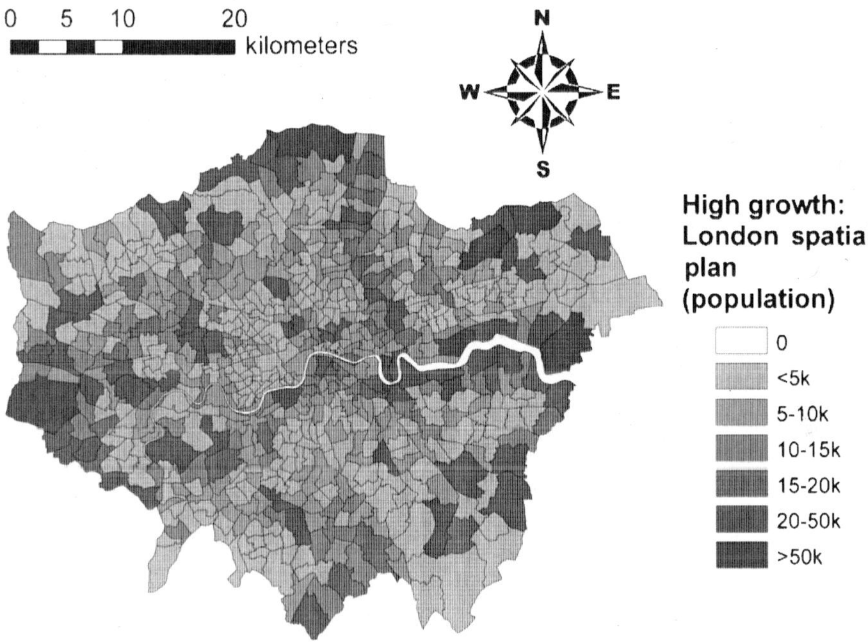

Sources: Trend-based employment forecasts for London by borough, Greater London Authority, http://www.london.gov.uk/mayor/economic_unit/docs/ep-technical-paper-1.pdf; current and historic population demographics, Office for National Statistics, http://www.statistics.gov.uk/default.asp; wards: census area statistics, U.K. Borders, http://edina.ac.uk/ukborders/; census data, U.K. Census Service, www.census.ac.uk; current land development, MasterMap, http://www.ordnancesurvey.co.uk/oswebsite/products/osmastermap/; index of deprivation, communities and local government, http://www.communities.gov.uk/communities/neighbourhoodrenewal/deprivation/deprivation07/; property type and location, National Property Database, The Environment Agency, http://www.environment-agency.gov.uk.
Note: To see this figure in color, refer to the appendix at the end of the book.

Analysis of each of these risks involves consideration of both the probability and consequences of harmful climate-related events. Scenarios of relevant climate variables are based on an existing Global Climate Model and regional climate model outputs, which are further downscaled as necessary. Analysis of flood risk is based on analysis of surge tides in the Thames estuary and flood flows in the river Thames. Forecasts suggest that surge tide frequency will increase as a result of projected changes in regional mean sea level (which is estimated to be in the range 0.19–0.88 meters in 2095). The potential for increasing frequency of cyclonic events that lead to surge tides in the southern North Sea has also been hypothesized in the literature (Lowe and Gregory 2005); however, recent analysis by the UK Met Office Hadley Centre indicates no significant trend in surge frequencies over the 21st century (Lowe et al. 2008).

FIGURE 3.7

Two Scenarios of Future Residential Development in 2020

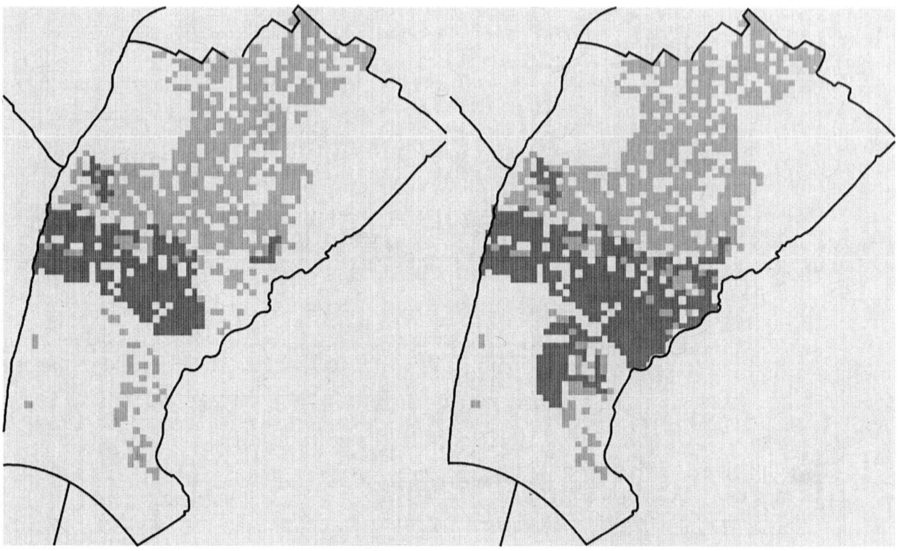

Sources: Trend-based employment forecasts for London by borough, Greater London Authority, http://www.london.gov.uk/mayor/economic_unit/docs/ep-technical-paper-1.pdf; current and historic population demographics, Office for National Statistics, http://www.statistics.gov.uk/default.asp; wards: census area statistics, U.K. Borders, http://edina.ac.uk/ukborders/; census data, U.K. census service, www.census.ac.uk; current land development, MasterMap, http://www.ordnancesurvey.co.uk/oswebsite/products/osmastermap/; index of deprivation, communities and local government, http://www.communities.gov.uk/communities/neighbourhoodrenewal/deprivation/deprivation07/; property type and location, National Property Database, The Environment Agency, http://www.environment-agency.gov.uk.

Note: Light gray indicates existing residential land cover; dark gray indicates future residential development.

The analysis of fluvial flood frequency has made use of downscaled regional climate model scenarios, propagated through a hydrological model of the Thames catchment. A similar approach was used to assess water availability under different scenarios of climate change. For the time being, flooding resulting from extreme rainfall within the urban area (pluvial flooding) and tidal flooding in the tributaries of the Thames have not been analyzed. Analysis of heat in the urban area is based on a combination of existing temperature measurements and a new version of the Met Office Hadley Centre's regional climate model, which includes the heating effects of the urban land surface (McCarthy, Best, and Betts 2009). The regional climate model is based on a 25-kilometer grid, but the subgridscale urban land surface scheme provides temperature outputs on a 5-kilometer grid. This enables the modeling of the effects of urban land cover and anthropogenic heat emissions. Figure 3.8 shows the pattern of maximum daily temperatures (1961–90 and in the 2050s), averaged over the summer (June, July, and August) season, which clearly illustrates the amplification of temperatures in central London and around Heathrow airport.

FIGURE 3.8
Daily Maximum Summer Temperature for London, 1961–90 and 2050

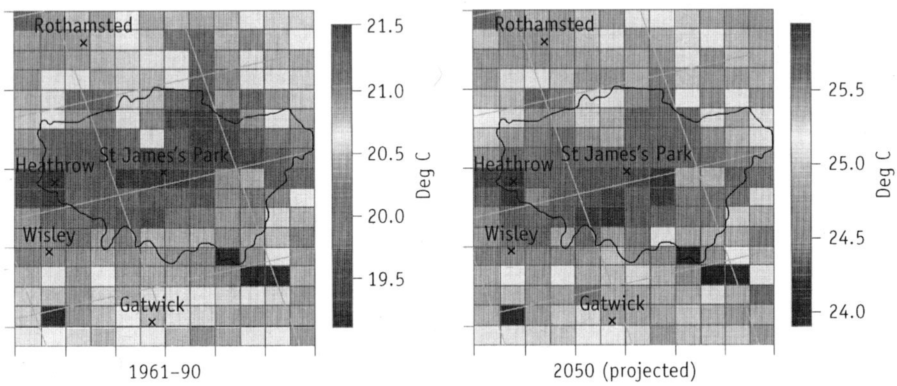

Source: Climate change projections, U.K. Climate Projections (UKCP09), http://ukcp09.defra.gov.uk/.
Note: Daily maximum summer temperatures, averaged over June, July, and August. To see this figure in color, refer to the appendix at the end of the book.

The impacts of climate change will often be felt in terms of the changing frequency of damaging extreme climate events. Therefore, climate impacts are typically measured in terms of changing average annual losses, which involves integrating over the extreme value distribution of the climate variables of interest. These distributions are combined with damage functions (for example, relating the depth of a flood to the duration of associated economic damage). These metrics of vulnerability will change in the future, as a result of changes in economic structure and patterns of land use; the economic and land use modeling described provides insights into these changes.

Figure 3.9 provides projections of the number of people at risk from tidal flooding under different development scenarios. The strong upward trend is driven primarily by changing levels of vulnerability and, to a lesser extent, sea level rise. The prevention of development in the floodplain will clearly be effective in limiting increased risks due to flooding, but it is probably an unrealistic solution because so much of the land available for development is located in the floodplains of London. Interestingly, the two planned development scenarios show a larger number of people at risk from floods than in the unconstrained development scenario. This is because of the emphasis in planned development on the redevelopment of brownfield sites, which in many instances are located in the floodplain.

The development of a platform to obtain future scenarios of climate impacts enables it to be used to test adaptation options as well. These options can include measures to reduce the impact of climate-related stresses, for example, by providing new water supply sources for the population before a water scarcity–related crisis situation has been reached, or changes to building design as a safeguard against extreme climate events.

FIGURE 3.9

Population at Risk of Tidal Flooding in London for Different Scenarios of Land Use Change

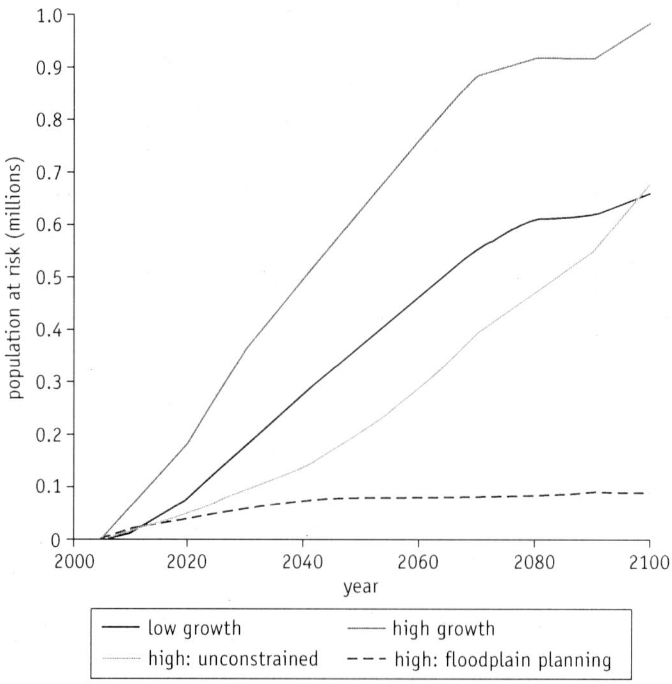

Sources: Current and historic population demographics, Office for National Statistics, http://www.statistics.gov.uk/default.asp; wards: census area statistics, U.K. Borders, http://edina.ac.uk/ukborders/; census data, U.K. census service, www.census.ac.uk; current land development, MasterMap, http://www.ordnancesurvey.co.uk/oswebsite/products/osmastermap/; flood defenses, Environment Agency's National Flood and Coastal Defence Database, http://www.environment-agency.gov.uk/static/documents/Research/protocol2_fr_apr03_1567934.doc; Topography, LiDAR, NextMap, IFSAR, http://www.intermap.com/nextmap-digital-mapping-program; floodplain areas, Environment Agency, http://www.environment-agency.gov.uk/homeandleisure/floods/31656.aspx; flood depth damage functions, *The Multi-Coloured Manual*, Environment Agency, http://www.environment-agency.gov.uk/.

Emissions Accounting

In addition to projecting climate impacts, the integrated assessment is also designed to provide projections of GHG emissions. The estimation of GHG emissions is based on the same demographic projections and outputs of the economic modeling described above. The emissions accounting tool associates GHG emissions with various levels of economic activity and population, based on a flexible set of variables that can be modified to reflect fuel mix, technological change, and energy demand. Figure 3.10 illustrates a typical projection based on a baseline economic growth scenario with no new mitigation policies. This effectively projects a business-as-usual scenario, which shows gradually increasing emissions, broadly in line with the changes in economic activity

FIGURE 3.10

Projections of GHG Emissions for London Based on a Baseline Economic Growth Scenario with No New Mitigation Policies

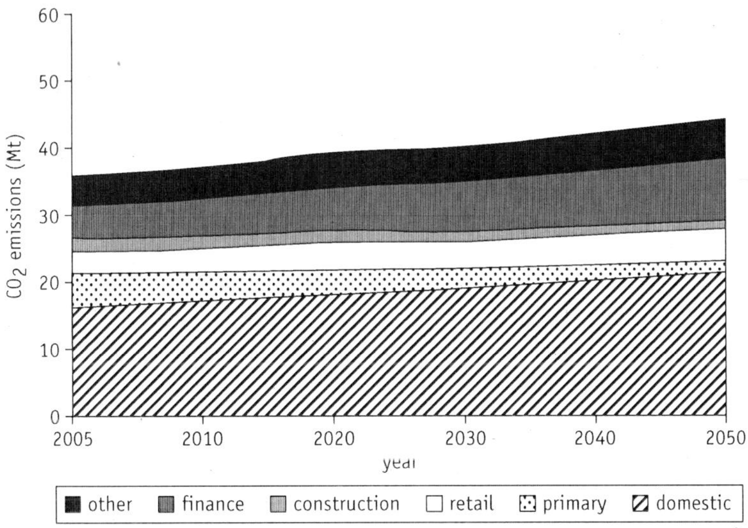

Sources: Atmospheric emissions in London, National Atmospheric Emissions Inventory, http://www.naei.org.uk; inventory of energy use, Department of Energy and Climate Change, http://www.decc.gov.uk/; energy statistics, *Digest of United Kingdom Energy Statistics*, http://www.berr.gov.uk/energy/statistics/publications/dukes/page45537.html; greenhouse gas inventory, U.K. Greenhouse Gas Inventory National System, http://www.ghgi.org.uk/unfccc.html; energy consumption statistics, Department of Energy and Climate Change, http://www.decc.gov.uk/en/content/cms/statistics/regional/regional.aspx; combined heat and power usage, The London Development Agency, http://www.lda.gov.uk/; on-site renewable energy database, The London Development Agency, http://www.lda.gov.uk/; energy use statistics, London Energy and CO_2 Inventory, http://www.london.gov.uk/gla/publications/environment.jsp; current and historic population demographics, Office for National Statistics, http://www.statistics.gov.uk/default.asp; future population demographics, exchange rates, interest rates, GDP growth, energy demand, national taxation, government expenditure, "Special Report on Emissions Scenarios (SRES)," http://www.ipcc.ch/.

projected in the MDM model (figure 3.3) The emissions estimates can be further disaggregated spatially based on existing and future patterns of spatial development.

Additionally, special modules have been developed for the generation of GHG emissions scenarios from personal and freight transport, using travel survey data as the basis for understanding existing travel patterns. Various policies in relation to infrastructure (loading bays, preferential lanes, consolidation centers, and capacity); regulation (delivery restrictions, road user charges, and low emission zones); technology (biofuels, hydrogen, and electrical vehicles); efficiency (driver behavior and delivery servicing plans); and modal switching have been tested. In general, these scenarios demonstrate that, while technological and efficiency measures can help to mitigate growth in GHG emissions, substantial reductions in GHG emissions are only achievable by constraining demand. The emissions accounting tool can be used to assess the potential effectiveness of energy efficiency measures at varying levels of uptake.

The accounting methodology has also been coupled with a scenario analysis tool to enable users to explore the impacts of different scenarios relating to energy demand, technology change, and portfolio options on modes of energy generation for energy supplies to cities. A key feature of the emissions accounting analysis is the ability to explore the cumulative emissions reductions necessary to achieve a given emissions reduction target.

Scenario Analysis

The descriptions of the various modules within the UIAF demonstrate that there are a large number of variables that may be specified by the user in order to generate scenarios of future change. These may be separated into exogenous variables over which city-scale decision makers in London have no control, or limited control, including climate change, demographic change, and rates of national economic growth, and decision variables, including those with respect to land use planning, transport, and other infrastructure planning and measures to improve energy efficiency. The exogenous variables in turn are developed into a set of scenario variables that comprise the following scenarios:

- A baseline economic growth scenario of 1.5 percent per annum, along with high (1.8 percent) and low (1.2 percent) annual growth scenarios.
- A flexible range of population scenarios for London.
- High, medium, and low climate change scenarios, corresponding to SRES A1FI, A1B and B1 scenarios (IPCC 2000), along with stable and increasing anthropogenic heat emissions from the urban area.

Decision variables are dealt with more flexibly than scenario variables, and the user may select a large number of combinations of different policies, implemented within a flexible range of timescales. Scenarios of transport connectivity, however, are much more time consuming to develop; hence, only three scenarios have been considered: the existing network (along with developments that are already under construction) alongside low and high long-term future transport infrastructure investment scenarios. Within the bounds of a given global scenario, national or citywide economic, transport, and land use policy can be tested, which does not necessarily have to coincide with the global scenario trajectory.

Scenarios of land use and city-scale climate and socioeconomic change inform the emissions accounting and climate impacts modules. Emissions accounting and climate impacts assessments are in turn informed by scenarios of economic and land use change, while being consistent with scenarios of climate change. Adaptation and mitigation scenarios developed within the integrated assessment framework must be consistent both internally and within the broader context of global change scenarios (for example, the technologies that may be adopted to mitigate transport emissions at a city scale cannot exceed the assumed level of technological advancement in the global scenario).

Support to Decision Making

The analysis described above has been brought together in an integrated assessment tool. This computer platform (currently implemented in Matlab[3]) holds all the variables that are communicated between the different modules. The more computationally efficient of these modules (for example, the spatial interaction module) are implemented within the central platform, whereas the more computationally demanding economic and climate analyses are run as offline models whose results are then uploaded into the integrating computer platform. The integrating platform is used to set scenarios, specify policy options, and store results.

The tool has enabled the research team to conveniently generate and display results as part of the testing and verification of the model. Figure 3.11 is a screenshot from the user interface that shows a scenario of land use change and a corresponding time series graph. The various buttons and drop-downs can be used to access maps and graphs of variables of interest.

Given the complexity of the analysis, the research team has sought to work interactively with stakeholders to identify and explore policy options. Specifically, the

FIGURE 3.11

Screenshot from the User Interface of the Assessment Tool

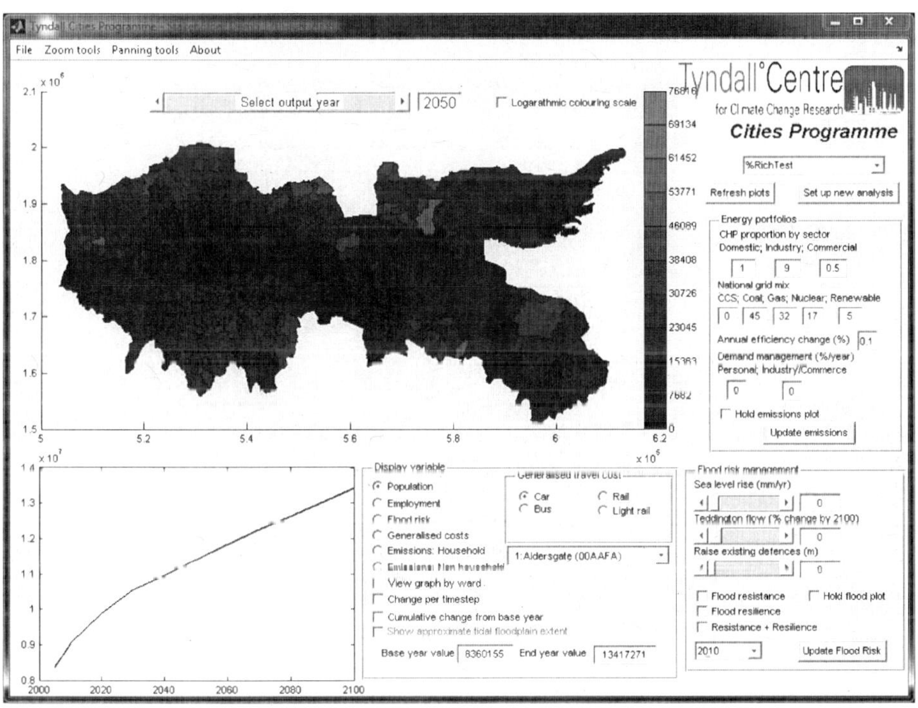

Source: Tyndall Cities Programme, Tyndall Centre for Climate Change Research.

range of transport options for the future scenarios mentioned above have been examined, and assessments of a range of land use planning policies that may be considered as part of the London Plan are a work in progress. The scheduled deliverables of the analysis are in the form of new quantified projections of the implications of alternative planning policies, which may be used to evaluate the impacts of the policies and identify desirable combinations of policy options.

Conclusions

The Tyndall Centre's Urban Integrated Assessment Facility (UIAF) brings together long-term projections of demography, economy, land use, climate impacts, and GHG emissions within a coherent assessment framework. It thereby provides the basis for examining at the scale of whole urban systems the effects of adaptation and mitigation decisions, with particular emphasis on decisions with an extended legacy. The UIAF is now being used to help inform decision making surrounding the new London Plan.

There are inevitable limitations to the number of process and interactions that can be included in a broadscale assessment of the type described here. The modules describing these processes are inevitably simplified. When brought together, they contain a very large number of scenarios and policy variables that may be set by the user. However, studying mitigation and adaptation at a systems scale provides the potential to understand systems interactions in a way that is not achievable in sector-specific assessments.

The implications of land use planning decisions in relation to climate vulnerability, in particular in relation to flood risks, have been illustrated in this chapter, along with brief discussions of the water resources and temperature assessment modules of the UIAF. Additionally, the GHG emissions assessment for London shows projections of steadily increasing emissions from industry, domestic, and transport sources unless strong measures are put in place to mitigate emissions. The feedback of climate impacts into the economy and land use have been targeted as priority areas for future research and are being addressed in the UK Engineering and Physical Sciences Research Council–funded project ARCADIA (Adaptation and Resilience in Cities: Analysis and Decision making using Integrated Assessment) now under way.

Notes

1. Vertical movement of the Earth's surface, which in the UK is associated with the melting of glacial ice masses.

2. Transport for London is the government body responsible for most aspects of the transport system in Greater London.

3. The Mathworks, http://www.mathworks.com/.

References

Barker, T., and W. Peterson, eds. 1987. *The Cambridge Multisectoral Dynamic Model of the British Economy*. Cambridge, UK: Cambridge University Press.

Environment Agency. 2007. "Water for the Future: Managing Water Resources in the South East of England—A Discussion Document." Environment Agency, Bristol, UK.

Evans, E., R. Ashley, J. Hall, E. Penning-Rowsell, et al. 2004. *Foresight. Future Flooding. Scientific Summary: Volume I—Future Risks and Their Drivers*. London: Office of Science and Technology.

GLA (Greater London Authority). 2004. *The London Plan*. London: GLA.

IPCC (United Nations Intergovernmental Panel on Climate Change). 2000. "Special Report on Emissions Scenarios." Intergovernmental Panel on Climate Change.

Junankar, S., O. Lofsnaes, and P. Summerton. 2007. "MDM-E3: A Short Technical Description." Cambridge Econometrics, Cambridge, UK.

Klein, R. J. T., E. L. Schipper, and S. Dessai. 2003. "Integrating Mitigation and Adaptation into Climate and Development Policy: Three Research Questions." Tyndall Centre for Climate Change Research, UK.

London Climate Change Partnership. 2002. "London Climate Change Partnership: A Climate Change Impacts in London, Evaluation Study, Final Report." London, UK.

Lowe, J., T. N. Reeder, R. Horsburgh, and V. Bell. 2008. "Climate Change Research and the TE2100 Project, Climate Change Impacts and Adaptation: Dangerous Rates of Change." Exeter, UK.

Lowe, J. A., and J. M. Gregory. 2005. "The Effects of Climate Change on Storm Surges Around the United Kingdom." *Philosophical Transactions of the Royal Society London*, 363 (1): 313–28.

McCarthy, M. P., M. J. Best, and R. A. Betts. 2009. "Cities Under a Changing Climate." Seventh International Conference on Urban Climate (ICUC-7), Yokohama, Japan.

McEvoy, D., S. Lindley, and J. Handley. 2006. "Adaptation and Mitigation In Urban Areas: Synergies and Conflicts." *Municipal Engineer* 159 (4): 185–91.

Najam, A., A. A. Rahman, S. Huq, and Y. Sokona. 2003. "Integrating Sustainable Development into the Fourth Assessment Report of the Intergovernmental Panel on Climate Change." *Climate Policy* 3 (S1): S9–S17.

ONS (Office for National Statistics) 2003. D. Virdee and T. Williams (eds.). *Focus on London*. Newport, UK: Office of National Statistics.

Pitt, M. 2008. "Learning Lessons from the 2007 Floods." Cabinet Office, London.

CHAPTER 4

Using an Integrated Assessment Model for Urban Development to Respond to Climate Change in Cities

Spike Boydell, Damien Giurco, Peter Rickwood, Garry Glazebrook, Michelle Zeibots, and Stuart White

This chapter describes an integrated assessment model for city-scale urban development that links the energy used in passenger transport (public and private) and residential in-house energy use. The model divides the urban region into disjoint subregions, the core of the model being centered on residential location choice, which is calibrated by population, demographic characteristics, and building types, leading to preferences for each subregion based on household type. Submodels are subsequently used to calibrate different rates of energy in accordance with household and demographic factors. This generates a picture of consumption patterns across the metropolitan area, enabling an appreciation of spatially heterogeneous factors such as differing levels of greenhouse gas (GHG) emissions, alongside variations in the distribution of infrastructures that can create considerable variation in energy consumption between districts within cities. The energy impacts of policy decisions that affect, by way of example, where new housing is to be built and of what type, can then be simulated. The workings of the model are demonstrated in the chapter using data on Sydney, Australia, as a case study, with the research offering a policy scenario to city officials to monitor its progress toward a 2030 vision for a sustainable Sydney.

Spike Boydell is a Professor at the School of the Built Environment and Foundation Director of the Asia-Pacific Centre for Complex Real Property Rights, Faculty of Design, Architecture, and Building, University of Technology, Sydney (Australia). Damien Giurco is Research Director for the Institute for Sustainable Futures, University of Technology, Sydney. Peter Rickwood is a Postdoctoral Fellow at the City Futures Institute, University of New South Wales (Australia). Garry Glazebrook is Senior Lecturer at the School of the Built Environment, Faculty of Design, Architecture, and Building, University of Technology, Sydney. Michelle Zeibots is a Senior Research Consultant for the Institute for Sustainable Futures, University of Technology, Sydney. Stuart White is a Professor and the Director for the Institute for Sustainable Futures, University of Technology, Sydney.

The overall objective of the model is to give policy makers a quantitative sense of how policy options (for example, changed land use policy or household efficiency improvement) will have an impact at the city scale. Thus, the spatially resolved outputs from the integrated assessment model provide a means of gauging the relative merits of different policy measures aimed at reducing GHG emissions in cities. This model is intended for use as a decision support tool by local, regional, and state government planning authorities, as well as energy and utility service providers.

Introduction and Objectives of the Model

This chapter describes a model for city-scale urban development that integrates urban residential water consumption, passenger transport, and in-house energy use in a single analysis platform. At present, the model focuses on residential in-house and transport-related energy use and water use. However, with further development, commercial and industrial consumption can also be included. The model has two primary objectives: (1) to provide policy decision makers with an assessment tool that can quickly gauge, in an integrated way, the relative impacts of different policy measures aimed at reducing water consumption, energy use, and greenhouse gas (GHG) emissions and (2) to enhance understanding of the material workings of cities by capturing the spatial interplay between infrastructure sectors, the demographic characteristics of the distribution of urban populations, and their consumption patterns.

With respect to the first objective, which relates to the effective administration of cities, the aim was to develop an integrated assessment tool for city-scale urban development inexpensively and quickly and using a minimum of data. The cost and time associated with acquiring and using the data are important because the responsibility for integrated analysis does not usually sit with any one particular agency. Consequently, such analysis is often overlooked and usually underfunded. However, in the interests of good governance, integrated analysis is an exercise that should be undertaken before decision makers commit to more detailed assessments of policy options.

The first objective recognizes that cities are complex systems whose management is overseen by a host of different government administrations and elected representatives. Each is responsible for the stewardship of different sets of infrastructure networks and services. The division of responsibilities into network types makes administration manageable. However, such management structures can mask the interplay between consumption patterns and the spatial distribution of infrastructure services that can significantly affect outcomes. Consequently, finding ways to integrate policy responses across infrastructure portfolios becomes an important function of government if sustainability objectives are to be achieved. In this context, the integrated urban model presented in this chapter offers analytical capability as a decision support tool for local, regional, and state authorities, and government departments. It also enables energy and utility service companies to model the impact of changed transport and land use configurations on consumption in relation to other impact mitigation measures.

The second objective of the model is to improve policy makers' understanding of the opportunities and potential barriers to achieving policy goals. Essential to this second objective is the recognition that the relative spatial location of different elements in the urban system affects consumption levels. For example, relative proximity to city centers clearly affects rates of energy use in the transport sector. Variation in local geography affects local temperatures and, hence, affects energy consumption for heating and cooling. How these factors interplay with the spatial distribution of demographic factors and building typologies is not always clear; however, their combined impact affects the degree to which energy usage and emissions can be reduced.

The model sets out to achieve the above objectives by first dividing the urban region into disjoint subregions. The core of the model then assigns building types and population and demographic characteristics to each subregion. It then uses submodels to assign different rates of energy and water consumption in accordance with household and demographic factors. This generates a picture of consumption patterns across the metropolitan area. This representation enables an appreciation of spatially heterogeneous factors such as differences in local climatic conditions alongside variations in the distribution of infrastructures and population demographics that can create considerable variation in consumption levels between districts within cities.

To develop the model, a transdisciplinary team of researchers was brought together with expertise in sustainability, climate change, urban resource management, transport planning, property theory, urban design, urban economics, spatial modeling, geographic information systems (GIS), and mathematics. The team's research results are presented in the next five sections. The second section provides a summary of the methodological approach used to compile the model and how it differs from prior research in the field. This includes an overview of the generic components of the model structure and how these can be adapted for use in different cities where data and administrative structures vary or where data may not be so readily available.

The third section describes the data used and explains the assumptions made to operate the model for an analysis of Sydney by the team. How the model might work in cases where it is assumed data are not so readily available, such as cities in developing economies, is discussed.

The fourth section presents the results of applying the model to Sydney. The model's capacity is demonstrated with a specific focus on energy use and greenhouse outputs. The analysis is expanded by focusing on housing demand and the implications of any underlying mismatches between building types, household composition, and income segregation, which demonstrates how the model produces outputs that stimulate a wider-ranging discussion about policy options rather than mere prediction.

The fifth section reflects on the results and articulates the lessons learned by the team to date. In particular, the practical difficulties in acquiring highly detailed consumption data are highlighted, together with a demonstration of how little bearing greater detail actually has on identifying general trends.

The sixth section makes recommendations on how further development/adaptation of the model might proceed, to enable its use in assisting decision makers responsible for the development and implementation of urban policies across key infrastructure and service sectors.

Methodological Approach to Compilation of the Model

This section discusses the methodological approach used to construct the integrated urban model. The discussion is presented in three parts.

The first part discusses recent *trends in urban modeling*, highlighting how the level of sophistication in modeling urban systems has paralleled the advancement of computing capability. This overview includes a review of urban model development in Australia, providing context for the approach and philosophy adopted in the overall discussion in this chapter of the integrated urban model.

The second part of this section discusses the *philosophy* that underscores the research team's decisions on how to structure the model. It explains how the objectives relating to institutional structures, data availability, and the complexity inherent in urban systems alluded to in the introduction were reconciled with technical aspects of modeling.

The third part discusses the *structural components of the model* and how they work. Also explained are how and why the model can be readily deployed in a variety of cities and why it can still provide useful insights into the likely outcomes from different policies in cases of limited data availability.

Trends in Urban Modeling

The level of sophistication in modeling urban systems has paralleled the advancement of computing capability. However, more complex models do not necessarily lead to more accurate models or better decision outcomes. What is required is a functional model embedded in effective decision-making processes involving researchers, policy makers, and citizens. The evolution of computational transport/land-use models (see Wegener 1994; US EPA 2000; Hunt, Kriger, and Miller 2005) has been summarized by Timmermans (2003) as three "waves" of development (see table 4.1).

Older land-use models, such as ITLUP/DRAM/EMPAL (Putnam 1983, 1991), investigate spatial interactions and remain in widespread use. In contrast, UrbanSim (Waddell 1998, 2002) takes a behavioral approach to capture complex interactions by predicting the behavioral ramifications of a particular policy scenario. At the development scale, UrbanSim models simulate decisions to build on undeveloped land in terms of the type and density of the development. Though the model has already had several applications, UrbanSim remains largely a work in progress, and the designers (Waddell and Borning 2004) acknowledge that many technical challenges remain in the context of modeling complex systems in urban areas.

TABLE 4.1
The Three Waves of Transport/Land-Use Models

Wave	*Type*	*Examples*
Wave 1	Aggregate Spatial Interaction Model	ITLUP (DRAM/EMPAL), LILT
Wave 2	Utility Maximizing Logit Models	UrbanSim, RELU-TRAN, TRANUS, MUSSA
Wave 3	Activity-Based Microsimulation Model	PUMA, ILUTE, RAMBLAS

Source: Adapted from Timmermans 2003.

Models such as MEPLAN (Echenique, Crowther, and Lindsay 1969; Echenique et al. 1990) and TRANUS (la Barra 1989) fit somewhere in between, relying on spatially aggregate economic interactions (derived from input/output tables) to determine general flows of goods and locational demand for labor. They engage interzonal flow information to determine location-specific demand for floorspace, rather than providing any explicit representation of firms.

Currently, there is a movement toward models that incorporate explicit interaction with businesses and households rather than using aggregate spatial interactions. Taking transport simulations as an example, within Timmermans (2003) "waves," the first wave treated travel behavior as a product of interacting spatial variables. In the second wave, the examination of travel is at the household level. Travel behavior is further deconstructed at the third wave, with household level behavior being broken down to the individual trip/activity.

The evolution of the three waves has seen increasing complexity in line with expanded computational capacity. This complexity comes at some cost in terms of applicability, portability, and intelligibility. Such models are time-consuming to develop and apply and often difficult to interpret. The trend in transport modeling—toward behavioral accuracy and away from intelligibility—while not unique to this area, illustrates the general issues involved in adopting complex computational models in a policy-driven environment.

The level of complexity increases when models are expanded beyond transport/land use simulations to integrate urban structure, building design, and domestic water consumption. Within the context of the particular research environment analyzed in this chapter (that is, metropolitan Sydney), there are several examples of mapping the relationship between water use and dwelling type (see Troy, Holloway, and Randolph 2005; Troy and Randolph 2006). These researchers assert that per capita water use in detached dwellings is similar to per capita consumption in apartments (apartments may also be referred to as units or flats in Australia), while detached dwellings (housing more people than apartments, and having a garden) use a greater volume of water per household. The New South Wales (NSW) government regulator (IPART 2004a) also found that detached households use more water than households in apartments. Studies in Australia seek further insight into water use behaviors by describing the water consumption down to the end use level, for example, toilet flushing, shower use, clothes washing, and garden watering (Roberts 2005; Willis et al. 2009).

While several studies have found a positive correlation between income and water use (for example, see Beatty, O'Brien, and Beatty 2006; IPART 2004a), the role of income as a driver for demand merits further research to understand the changes in end uses that lead to this finding. The need to consider the energy implications of urban water supply should also be emphasized—particularly in the Australian context where persistent drought has necessitated the construction of desalination plants to augment supply and encouraged the significant uptake of rainwater tanks (with associated energy costs for pumping) in homes (Retamal et al. 2009).

An urban Australian study by Randolph and Troy (2007) explores the extent to which dwelling type and the socio-behavioral characteristics of households influence the pattern of electricity and gas consumption. The study provides useful insights on household practices and attitudes toward energy consumption, with notable differences between house dwellers and apartment dwellers and further variations between low-rise and high-rise unit dwellers. However, Randolph and Troy were unable to link actual energy consumption data with individual survey data because of data protection and privacy legislation.

Attempts to analyze interdependencies between urban structure and energy use are fraught with problems. First, data required for such a meta-analysis remains fragmented, and access to linked data raises privacy problems. Second, many analyses fail to provide appropriate comparisons. Several studies have been undertaken in this genre. Myors, O'Leary, and Helstroom (2005) compared recently built high-rise apartments and housing stock in general and often found higher levels of energy consumption in high-rise apartments.

A Canadian study by Norman, MacLean, and Kennedy (2006) compared high- and low-rise residential density to provide an empirical assessment of energy use and GHG emissions arising from transport, operational energy, and building construction (including embodied energy). They found that energy use in low-density suburban development was twice as intensive as in high-density development on a per capita basis. Studies have also shown that smaller houses can appear more energy intensive if only assessed on a unit area basis without taking into account house size, number of occupants, and total energy used (Thomas and Thomas 2000).

Three other examples of urban models relevant to the Sydney case study are presented in this chapter. The first is the Sydney Strategic Transport Model (TDC 2007), an analysis of disaggregated transport and traffic patterns that draws on the census of population and housing every five years (ABS 2006). Based on a moving sample of some 4,000 households, it includes detailed sociodemographic data, journey-to-work data, and a continuous Household Travel Survey (HTS). The second is the Melbourne Region Stocks and Flows Framework (MRSFF), which integrates a range of different models to analyze the "city metabolism" in order to characterize the interactions between model components, such as buildings and demography, within the "what-If?" modeling environment.[1] The outputs are forecasts of development over short-, medium-, and long-term time horizons (Baynes, Turner, and West 2005). The model is distinguished by the big picture, aggregated level analysis of the main development patterns it provides as output.

Finally, BASIX—the online Building Sustainability Index introduced by the New South Wales government[2]—is a compulsory assessment tool created to ensure that new homes are designed to use a lesser amount of potable water (40 percent reduction target) and be responsible for less (25 percent reduction) greenhouse gas emissions. A critical component of the policy tool is the database for each application, which includes information on location, house size, and building design and also includes measures for energy and water efficiency. From the perspective of this research, the database is constrained in that it only contains information regarding buildings where development consent has been granted over the last five years.

From the above examples, it is evident that a significant amount of data is required to calibrate the more sophisticated transport/land-use models, and this has been an impediment to their widespread adoption. The more ambitious the scope of a model and the more effort put into modeling all the factors that influence household and firm decisions, the more complex the model becomes and the more data are required calibrating that model.

Modern models based on behavior at the household/firm level are seen as superior because they describe the urban systems being studied more accurately, and the trend among the urban-modeling research community is toward greater levels of sophistication and detail. However, accurate input data on travel time and cost are often many years out of date, and when combined with (differently) out-of-date data on land prices, employment distribution, and fuel price elasticity, the validity of the output is often undermined.

While a simple model can offer some approximation of reality, the tendency to proceed to refining the model can be a distracting journey toward a hypothetically "true" model. Models can only approximate reality by providing a useful mental tool rather than a faithful representation of truth. There is a point of diminishing returns, as a model grows more complex. Our intention in developing an integrated urban model was to remain focused on the broader role the model plays in informing effective decision-making processes.

Philosophical Approach to Modeling Urban Systems

As outlined in the introduction, cities are complex systems. Complexity occurs for several reasons, including the high degree of stochastic uncertainty in the behavior that takes place in urban networks and because there is an interaction between multiple subsystems (Baynes 2009). Despite the complexity of urban systems, comparative empirical studies have shown that behavior also conforms to distinct patterns in accordance with broad generic organizing principles (for example, Thomson 1977; Kenworthy and Laube 1999).

With these conditions in mind, we approached the compilation of the model by accepting that urban land use and transport models will always suffer from several limitations. One of these factors is the limits in the precision that is possible in rendering consumption patterns and problems with long-range forecasting, given the fundamental uncertainty in both the administration and material workings of cities.

Faced with these limitations, it is reasonable to view land use and transport models as better for exploring different scenarios and policy options than for making long-range forecasts. Timmermans (2003) articulates this view in some detail, suggesting that expectations and claims of models need to be adjusted, acknowledging that they can provide a useful indication of the direction in which behavior and consumption might pan out given certain conditions. However, they are unlikely to provide highly accurate quantitative assessments of urban behavior over the long term.

From this perspective, using models for scenario evaluation rather than forecasting can assist the planning process by helping decision makers and the wider community realize what some of the general outcomes from decisions could be. By presenting different possibilities, a model used in this way encourages dialogue between different service providers and a greater exploration of possible policy options. In contrast, when the focus is on the use of very complex models that provide *precise* answers, debate and exploration of ideas can be stifled because model outputs tend to be treated as facts rather than useful, but fallible, explorations of what is possible (Timmermans 2003).

With acceptance of the limitations inherent in the modeling process in conjunction with the limitations that government agencies have in relation to resource allocation, the model this research team is developing sits somewhere between those very complex models at the forefront of land-use and transport modeling and simpler econometric/statistical models. This complexity level ensures that data-intensiveness does not present a barrier to considering the impacts of climate change response policies at the city scale.

Structural Features of the Integrated Urban Model

The structural features of the integrated urban model comprise a *residential location choice model* at the core of the model structure (see figure 4.1) with a range of submodels that calculate corresponding energy consumption. The unique feature of the residential location choice model is that it does not require data on housing prices in order to be calibrated.

The components of figure 4.1 describe the model in two phases: calibration and exploring the impact of policy decisions. The calibration phase requires current data (or recent historical data from the latest census year) as an input. Precise data requirements are detailed further in the next section; however, in broad terms, these data relate to

- household type (how many people live in the house, are they young or old, single or married, with or without children?),[3]
- household income level,
- dwelling type (larger detached house, smaller terrace house/townhouse, or apartment/unit), and
- location.

FIGURE 4.1

Integrated Model Concept

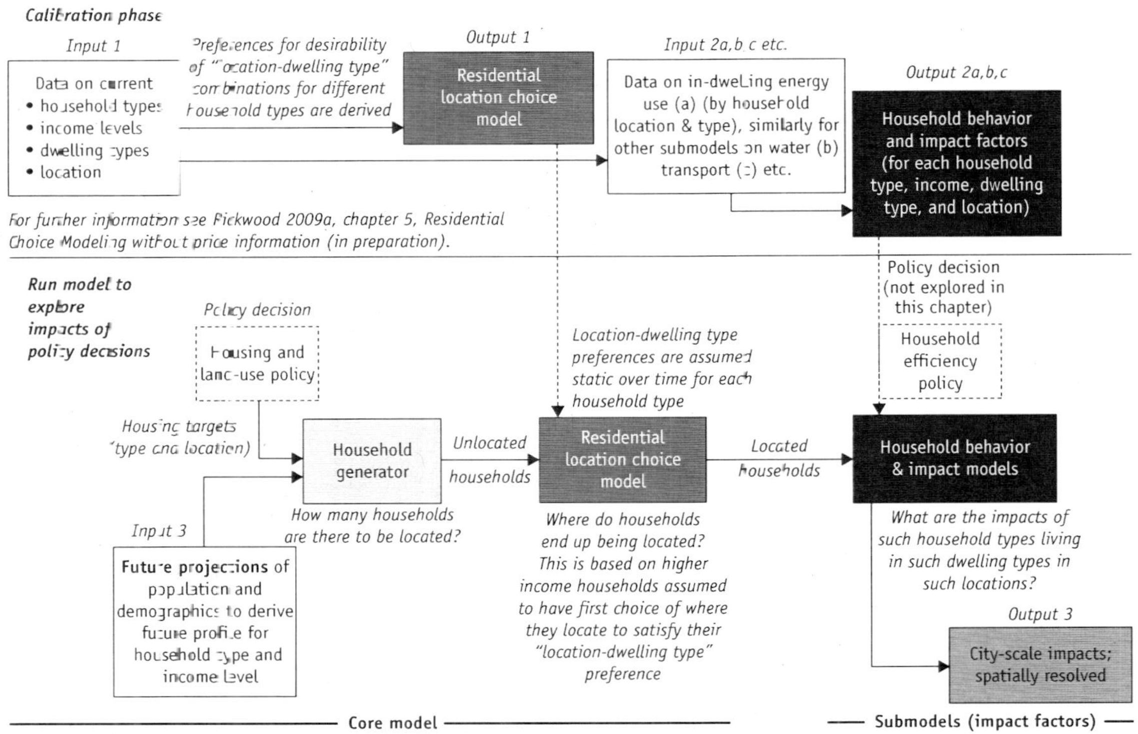

Source: Adapted from Rickwood et al. 2007.

By examining where people in different household types live, the user constructs a "preference model" for each household type, where each has its most desired "location-dwelling type" combination; this assumes that higher-income households get to realize their preferred choice. This set of preferences is what underpins the residential location choice model (Output 1) and is assumed to apply in future policy scenarios explored later. Therefore, the model does not take into explicit account some suburbs becoming "trendier" and thus more sought after for specific or all household types.

Input 2 shows the need for spatially resolved data on actual consumption (for example, in-dwelling energy, in-dwelling water, transport energy), which is used to calculate impact factors for households based on their type, income, dwelling, and location. This calculation leads to derivation of the submodels for household behavior and impacts and completes the calibration phase.

The model is then run in the second phase to explore impacts of future policy decisions, represented in dashed boxes in figure 4.1. Here it can be seen that housing and land-use policy would affect the household generator component of the model. In contrast, a policy aimed at increasing household energy efficiency by subsidizing roof insulation would affect the household behavior and impact factors.[4]

Expanding the housing and land-use policy would set future targets for household types and locations for new residential dwellings. In addition to the policy projection, Input 3 is required, which comprises future projections of population and demographics (from which future profiles for household type and income level are derived). Together, this information supplies the household generator component of the model to determine how many new households must be established under the policy. These unlocated households are then sorted using the residential location choice model (from the calibration phase) to assign households to location, according to a queuing approach in which higher-income households choose their preferred location-dwelling type combination first. This gives rise to fully located households whose impacts (for example, energy) are then tallied based on the calibrated household behavior and impact models.[5]

The above approach differs from the usual approach used in urban models, where household and firm submodel structure is specified as part of the model. The separation into core and submodels is deliberate so that the model can be widely used by being flexible enough to accommodate variations in data availability across different cities. In some cities, access to rich disaggregate datasets will allow complex firm and household behavior submodels to be attached to the core model. In other cities, however, development of such submodels may not be possible. For example, rates for consumption might be estimated by comparison with other cities that have comparable conditions. Irrespective of local data availability, the key point is that by separating the model core from the model subcomponents, it is possible to tailor submodels to the data available in each city.

To illustrate, two models are analyzed—a travel model and a dwelling-related energy model, which will be discussed in the results section. It is important to note that the number and nature of the submodels attached can be varied depending on the circumstances. For example, if data were not available to develop a disaggregate travel model, there would be nothing to prevent an aggregate spatial-interaction style travel model being attached instead. Alternatively, if policy makers were interested in matters other than transport and in-dwelling energy use, then other components that model the behavior of interest could be incorporated. As a result, the amount of data required to calibrate the entire model is largely under the control of the modeler. By selective simplification, a model has been developed that is complex enough to model household behavior at a fine spatial scale, while ensuring that it is easily portable/applicable to just about any urban area.

Data Used and Assumptions Made to Operate the Model

This section outlines further key data and assumptions for conceptual model described in the previous section. The data types and sources used for actual data in the Sydney case study are shown in table 4.2. Considering the modeling approach in a more general sense, the categories used in the Sydney case study—for example, with respect to household type (eight data categories were used) or income level (six data categories)—could be changed according to available data and required groupings for policy analysis.

The key assumptions for the model are summarized in table 4.3. The limitations of these assumptions for the Sydney case study, and in a more general sense with a view to application in other cities in the developing and developed world, are discussed in the fifth section, "Limitations of the Model."

Having explored the data and assumptions used in this model, the next section applies the model to a case study based in Sydney, with a focus on transport and in-dwelling energy (for residential homes; commercial and industrial premises are not included).

Results to Date from a Sample Analysis of Model Outputs for Sydney, 2006–31

The authors, who are based in Sydney, were initially funded to test the capacity of the model in the context of the Sydney metropolitan region. In 2005, the Sydney Metropolitan Area contained some 4.2 million people. The NSW Strategy for Sydney anticipates the population of metropolitan Sydney will increase to 5.3 million by 2031 (an increase of 1.1 million people between 2004 and 2031) (NSW Department of Planning 2005). An overview of Sydney in the context of its corresponding metropolitan strategy is provided in figure 4.2. While Sydney has added 1 million residents since 1975, its water consumption remains the same due to more efficient water use. Australia has the highest per capita GHG emission rate of any developed nation, with each person in Sydney currently creating 27.2 tonnes of CO_2 per annum.

TABLE 4.2
Types of Data Used in the Integrated Model

Residential choice location model	*Source*
Household type	2006 census data (Australian Bureau of Statistics)
Young single occupant	
Old single occupant	
Young couple	
Old couple	
Single parent	
Couple with children under 15	
Couple with children over 15	
Other	
Income level (weekly income, AUD)	2006 census data (Australian Bureau of Statistics)
Under $650	
$650–$999	
$1,000–$1,399	
$1,400-$1,999	
$2,000-$2,999	
$3,000+	
Dwelling type	2006 census data (Australian Bureau of Statistics)
Detached house (larger)	
Townhouse/semi-detached (smaller)	
Apartment/unit (assumes no garden)	
Location	2006 census data (Australian Bureau of Statistics)
615 regions within the city, using a combination of areas represented in the census and travel survey	Household travel survey (NSW Transport population data centre)
Household behavior and impact factors	
In-dwelling operational energy use (electricity, gas)	NSW Independent Pricing and Regulatory Authority Survey of 2,600 homes in different regions
Travel energy use	Calculated from household travel survey (trip duration and frequency and mode) and other transport data
Embodied energy use in building types	Using a first principles differentiation

Sources: ABS 2001, 2006; IPART 2004a, 2004b, 2006; TDC 2007.

Single and two-person households are in the majority in Sydney, with 22 percent of households comprising one person; this number is anticipated to increase to 30 percent by 2031, requiring an additional 300,000 single-person households. Meanwhile, forecasts suggest households with couples and children will increase by 140,000 over the same period (NSW Department of Planning 2005, pp. 24–29).

The model, in the context of the Sydney case study, initially incorporates energy and GHG emissions data. Subsequently, aspects of housing satisfaction and income segregation are integrated to show how the model produces outputs that have the

TABLE 4.3

Key Assumptions Used in the Integrated Model

Key assumptions	*Comment*
Residential choice location model	
Assumes preferences are static over time	This does not specifically allow for suburbs becoming more desirable over time.
Assumes established transport routes are static over time	For example, the impact of establishing a new train line would require further input recalibration to be used in the model.
Assumes higher-income households can satisfy their preferences first	This is a key feature of the model, reducing the data required for the model; specifically no data on house prices are needed.
Household Behavior and Impact Model	
Assumes climatic conditions similar in regions of the city in future	As climatic effect (that is, hotter inland suburbs use more energy for air conditioning than seaside suburbs) is accounted for indirectly in the observed data patterns, future changes to which areas of the city are hot or cold are not explicitly included.
Greenhouse impacts	Average greenhouse factors for each transport mode and distances travelled.

Source: Rickwood 2009a.

FIGURE 4.2

Sydney in the Context of the Metropolitan Strategy

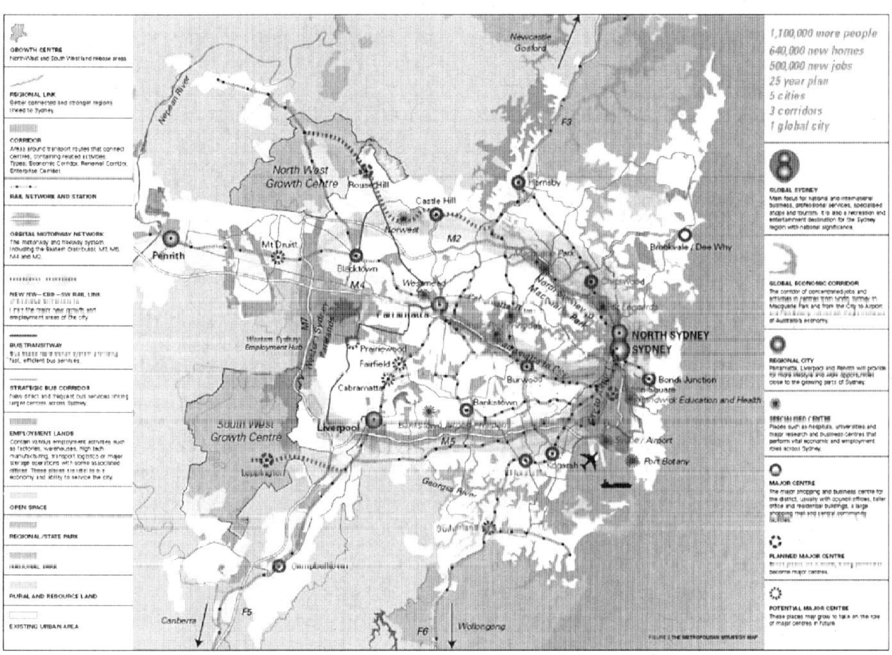

Source: NSW Department of Planning 2005, 10–11. Reproduced with kind permission of New South Wales Government Department of Planning.

Note: To see this figure in color, refer to the appendix at the end of the book.

capacity to stimulate a wider range of options, rather than relying on a prediction of what might happen. The outputs, as presented in this chapter, are only for the baseline scenario. The baseline scenario is grounded on the Sydney Metropolitan Planning Strategy (NSW Department of Planning 2005). Land use is exogenously determined to reflect policy decisions (that is, the user provides it as an input, in this case based on Metropolitan Planning Strategy forecasts).

Figure 4.3 shows the projections for new housing throughout the Sydney Metropolitan Area up to 2031 (a color version is provided at the end of the book). Dark blue indicates areas where less than one additional dwelling per hectare is proposed to be built by 2031, while those districts shown in red are anticipated to host increases of more than 50 dwellings per hectare, given the housing and land-use development

FIGURE 4.3

Sydney Exogenous Housing Inputs: New Dwellings per Hectare, 2006–31

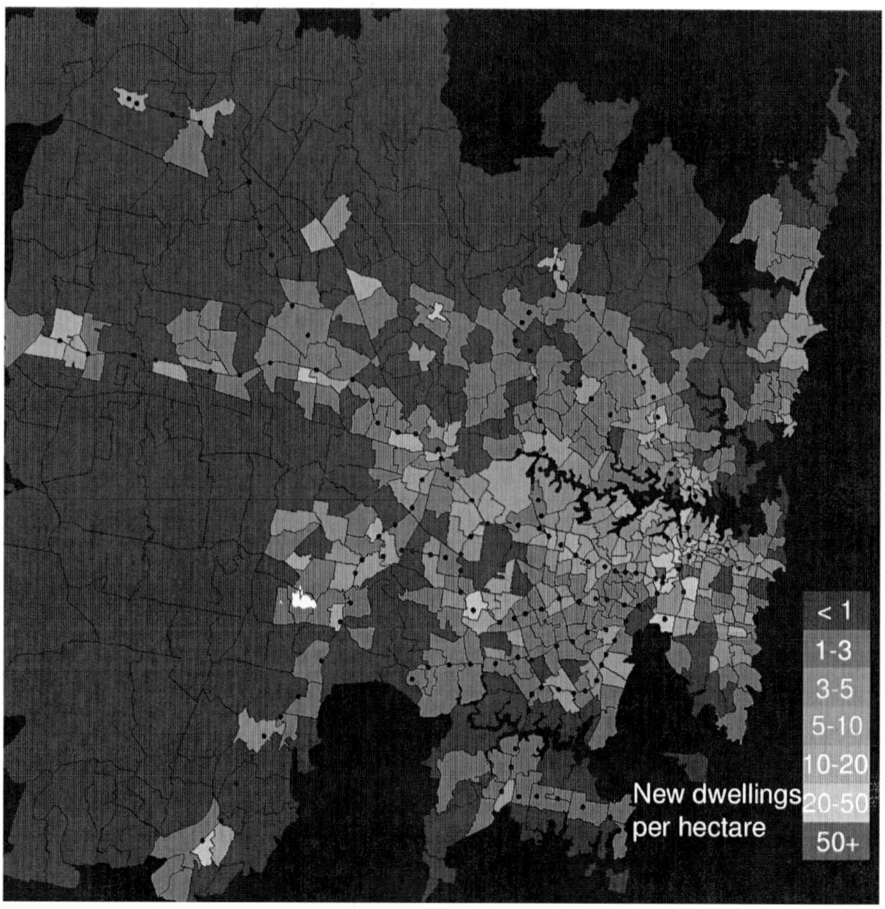

Source: Authors, based on Rickwood 2009a, 252.
Note: To see this figure in color, refer to the appendix at the end of the book.

policies specified in the Sydney Metropolitan Planning Strategy (NSW Department of Planning 2005).

From a GHG reduction perspective, the areas shown in green are of particular interest because they indicate where the largest proportion of new dwellings will be located in terms of total numbers. The policy includes a mix of development in existing urbanized areas—or *brownfield* sites—and new developments that would extend beyond the existing urban boundary—or *greenfield* sites.

Spatial variation in per household and per capita income is shown in figures 4.4 and 4.5. As can be seen, income levels per household are shown for 2001 in figure 4.4(a) based on census data, while the projected location for households in 2031, calculated by the residential location choice model, are shown in figure 4.4(b). Figure 4.5 shows the location of people based on income per capita for 2031, as projected by the residential housing choice model.

The spatial distribution of income per capita is important because it changes GHG emissions levels in ways that might be unexpected, providing significant insights for policy makers as to the effectiveness of their policies.

In the following series of outputs from the model (figures 4.6 and 4.7), the likely implications of the proposed housing policy are examined in terms of energy consumption and GHG emissions.

Figure 4.6 shows projections for the total dwelling-related primary energy used by households for 2031. This includes energy used within the home for heating, cooling, lighting, and other household appliances and tools. It also includes energy embodied in residential dwellings. As can be seen in figure 4.6(a), when calculated on a household basis, those dwellings located on the fringes of the urbanized area toward the west are higher consumers than those located closer to the central business district (CBD) to the east of the metropolitan area. However, when calculated on a per capita basis, as shown in figure 4.6(b), energy use is much higher for those living relatively close to the city center.

While the distribution of household types is not shown, the reason for this difference is a higher proportion of single-occupant and higher-income households locating in the areas within close proximity to the CBD, while those households living on the outer fringes have a far higher proportion of couples with children. A combination of factors are at play that include higher-income earners being able to use more energy per capita and multiperson households making per capita savings because they share appliances and living spaces.

Figure 4.7 shows projections for the total transport-related primary energy used by households in Sydney in 2031. This includes private passenger travel and public transport use. Unlike in figure 4.6, there is little difference in the spatial pattern of transport energy use when measured on a per household or per capita basis. This is because transport energy use, compared to in-dwelling energy use, is primarily influenced by relative proximity to public transport. Or, in other words, people—regardless of household structure or income level—will use lower energy public transport modes if they live close to them. Those districts with higher per capita transport-related energy use do not have rail and high-quality public transport service provision. If the

FIGURE 4.4
Sydney Household Income Deciles in 2001 and 2031

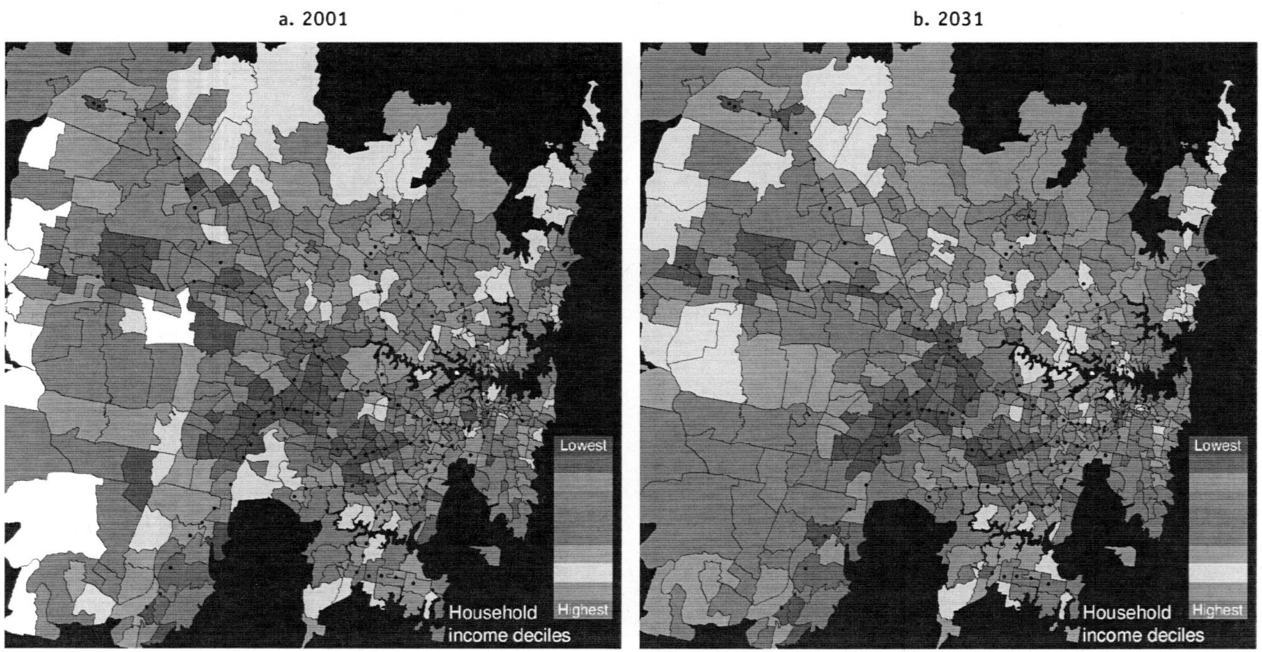

Sources: Authors, based on ABS 2001, 2006; Rickwood 2009a, 271.
Note: Household income deciles in (a) 2001 from ABS data and (b) 2031 projected for baseline scenario. To see this figure in color, refer to the appendix at the end of the book.

FIGURE 4.5
Sydney per Capita Income Deciles in 2031

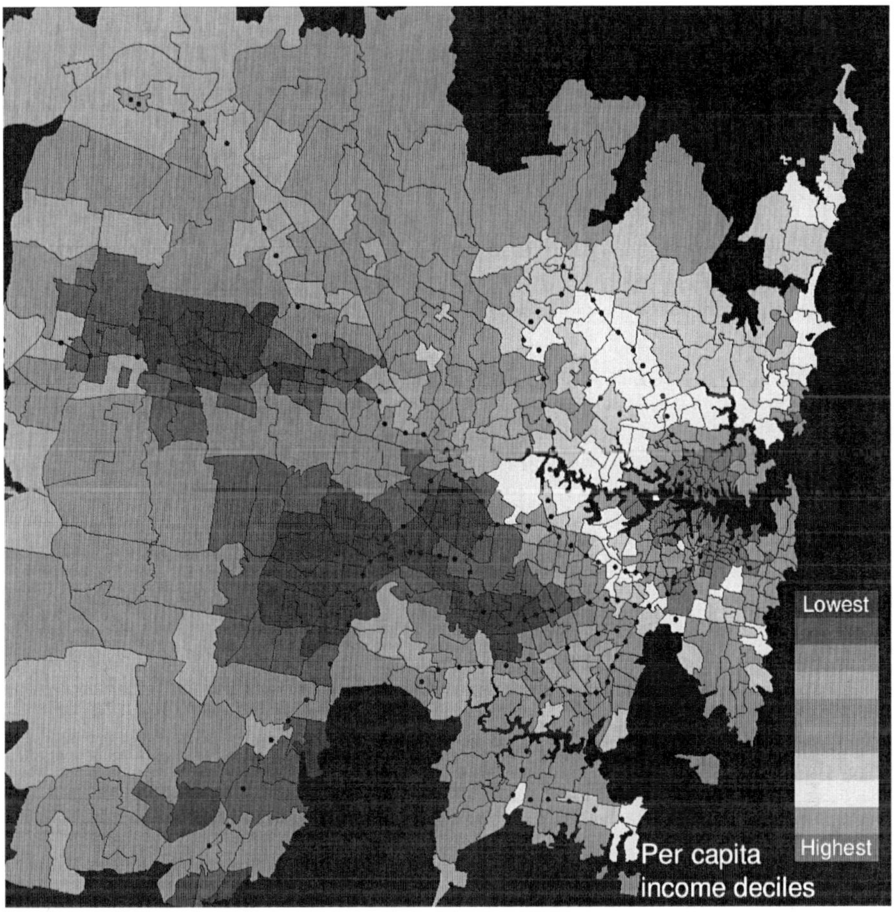

Source: Authors, based on ABS 2006 data.
Note: Per capita income deciles in 2031 (projected for baseline scenario). To see this figure in color, refer to the appendix at the end of the book.

location of the Sydney rail network and rail stations were not shown on the model outputs, it would be possible to pick the location of corridors served by rail simply by looking for differences in transport-related energy use.

The policy lessons that arise from these projections can be appreciated when the following outputs from the model (figures 4.8, 4.9, and 4.10) are taken into account.

Figures 4.8 to 4.10 show the spatial pattern of per capita GHG emissions for the 2031 projections. Figure 4.8 shows the emissions resulting from dwelling-related energy use, while figure 4.9 shows the emissions from residential-transport-related energy use. Figure 4.10 shows the combined emissions, or total GHG emissions, per person. Significantly, figure 4.10 shows that transport-related emissions dominate the overall spatial distribution of emissions in Sydney.

FIGURE 4.6

Sydney Annual Dwelling-Related Energy Use by Zone in 2031

a. Per household

b. Per person

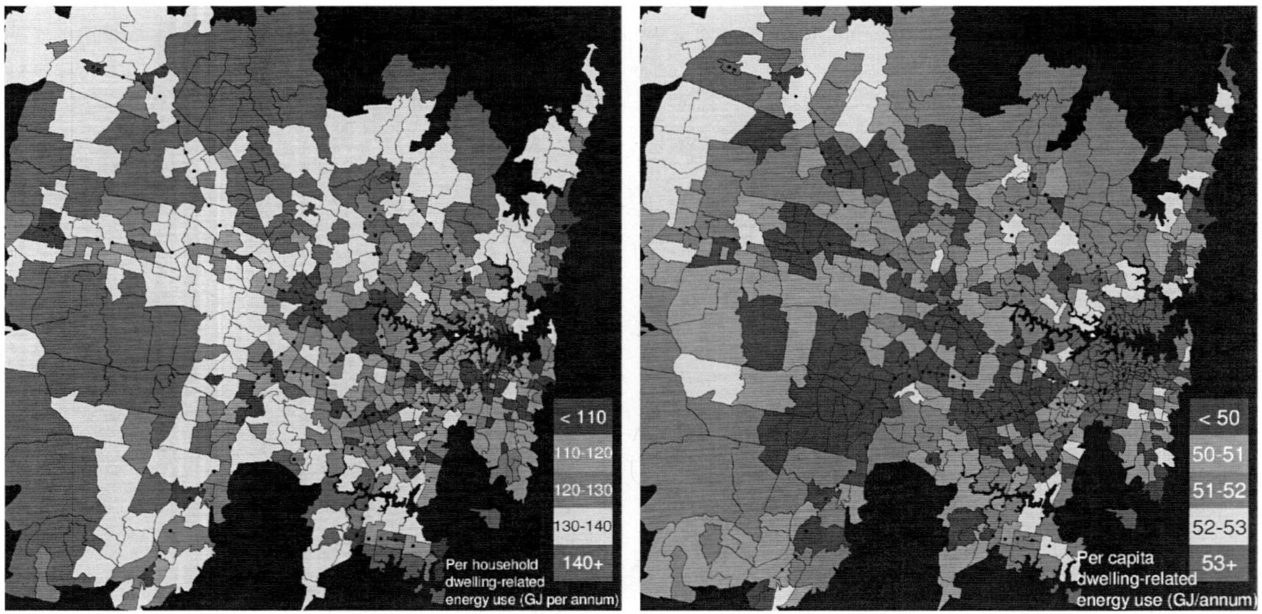

Sources: Authors, based on ABS 2001, 2006; IPART 2004a, 2006; Rickwood 2009a, 2009b.

Note: Includes embodied energy. To see this figure in color, refer to the appendix at the end of the book.

FIGURE 4.7

Sydney Annual Personal-Transport-Related Energy Use by Zone in 2031

a. Per household b. Per person

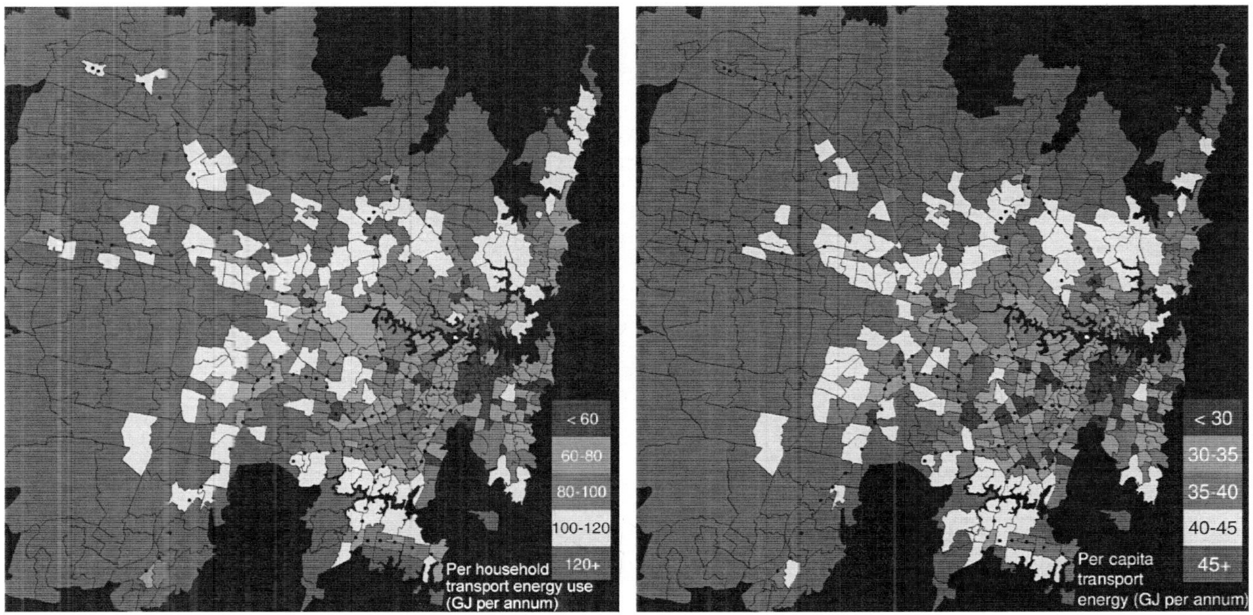

Sources: Authors, based on ABS 2001, 2006; Rickwood 2009a; TDC 2007.

Note: Includes energy embodied in cars. To see this figure in color, refer to the appendix at the end of the book.

FIGURE 4.8

Sydney Annual Dwelling-Related Emissions per Person by Zone in 2031

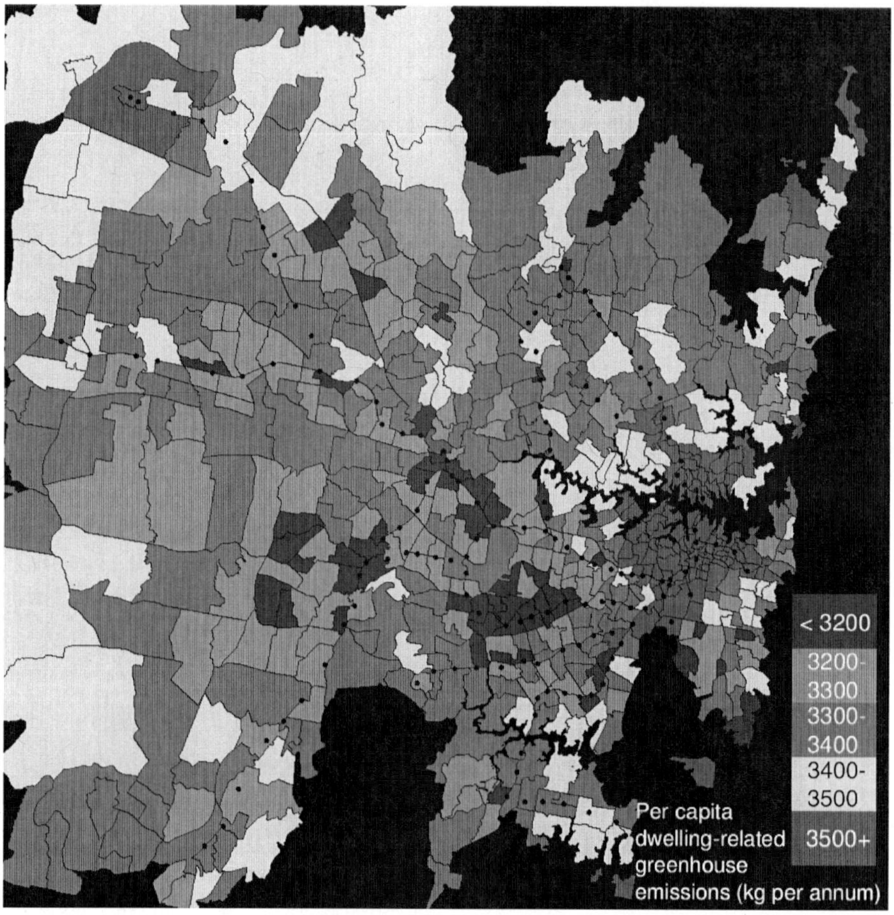

Sources: Authors, based on ABS 2001, 2006; IPART 2004a, 2006; Rickwood 2009a, 2009b.
Note: Annual dwelling-related emissions per person includes embodied energy. To see this figure in color, refer to the appendix at the end of the book.

The significant point arising from this analysis is the importance of public transport provision in reducing GHG emissions from urban populations. At present, policy makers and city officials have placed a great deal of emphasis on *green building* programs to bring down emission levels. However, emission reductions from the existing stock of buildings must also be tackled for the city to adequately respond to climate change. Furthermore, the analysis produced by the integrated urban assessment model shows that the provision—or lack thereof—of public transport services has a pervasive impact on GHG emission levels.

In recent decades, transport policy in Sydney has favored road development over investment in public transport. From the insights provided by the model, it would appear that policy makers should pay greater attention to transport policy,

FIGURE 4.9
Sydney Annual Personal-Transport-Related Emissions per Person by Zone in 2031

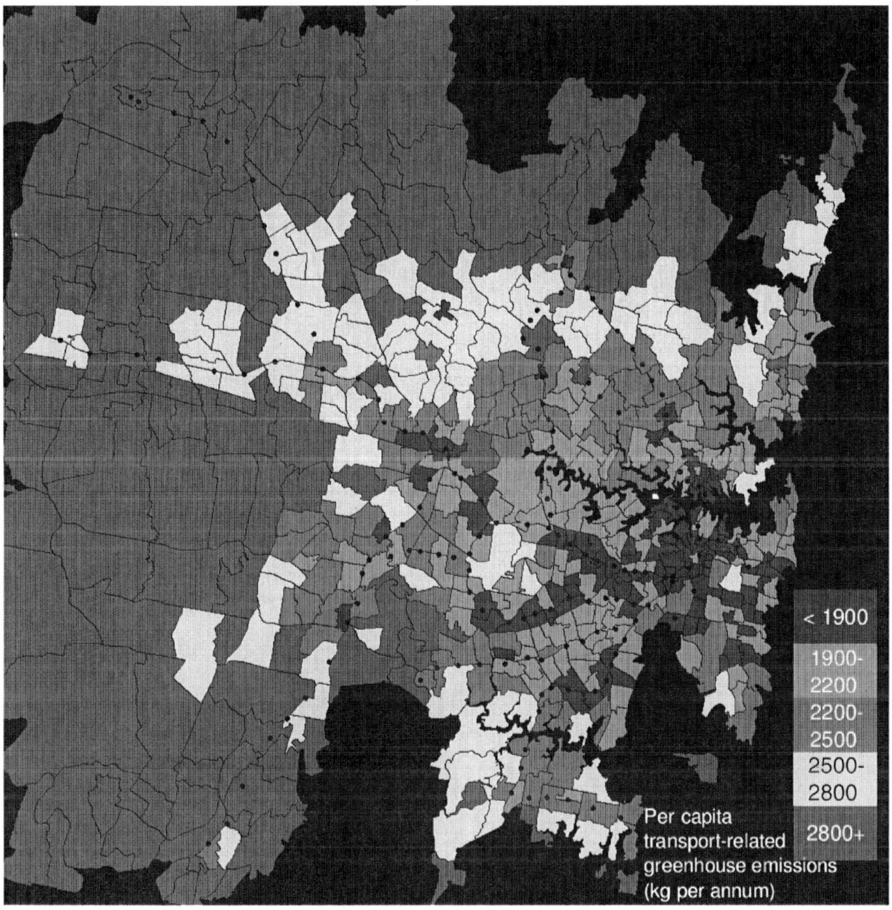

Sources: Authors, based on ABS 2001, 2006; Rickwood 2009a; TDC 2007.
Note: Includes emissions embodied in cars. To see this figure in color, refer to the appendix at the end of the book.

in particular on extensions to the public transport network, especially in those areas that currently have inadequate public transport systems. In the context of the city of Sydney analyzed in these models, this would constitute a significant shift in transport policy

Limitations of the Model

Accepting the limitations inherent in the modeling process, the research team chose to develop a model that occupies a middle ground between the very complex models at the forefront of transport/land-use modeling research, and simpler

FIGURE 4.10
Sydney Annual Emissions per Person by Zone in 2031

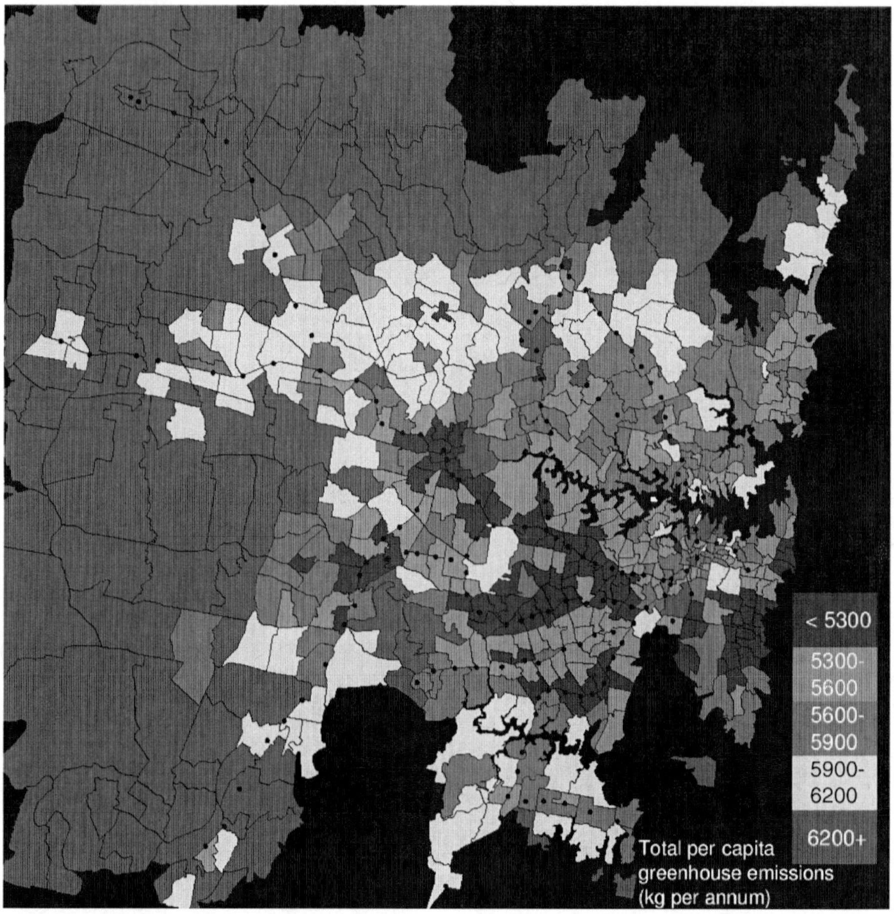

Sources: Authors, based on ABS 2001, 2006; IPART 2004a, 2006; Rickwood 2009a; TDC 2007.
Note: Includes emissions embodied in cars. To see this figure in color, refer to the appendix at the end of the book.

econometric/statistical models. A key benefit of this approach is that the model has relatively modest data requirements and, hence, has the potential for application in other cities. Also, despite being somewhat simpler than other recently developed models, the model developed is sophisticated enough to generate a rich set of visual and other outputs to usefully facilitate decision making.

The authors intended that the level of detail necessary to develop the model be sufficient for purposes of policy-related decision making. For example, focusing on green buildings alone may not be enough to bring emissions down to the levels required under targets determined for the city. It was accepted that lower-impact household types (for example, higher density dwellings) may not see the reductions in energy that were expected, given that there is evidence (Myors, O'Leary, and Helstroom 2005) that occupants with higher income levels who inhabit these dwelling types may consume more energy for appliance usage.

The integrated urban model presented in this chapter was funded by the University of Technology, Sydney, as a transdisciplinary research collaboration to analyze the prevailing and anticipated situation in the context of a city, Sydney, in a developed economy, Australia. The research team was supported by the utility providers and the NSW government in its data collection efforts. However, given the relative simplicity of the model inputs, the authors contend that data requirements would not reduce the portability or adaptability of the model to a developing country context.

Recommendations and Conclusions

This chapter outlined the structure and function of an integrated model developed to understand how land-use planning policy affects energy and transport. The development of such a model addresses a key deficiency with respect to planning for efficient, resilient cities—namely, the lack of an integrated platform for analysis at the city scale. Demonstrating the application of the model to the city of Sydney led to the following conclusions:

- Energy use related to personal transport is lower, per person *and* per household, in the city center and along public transport corridors than in the outer suburbs where car-based travel dominates trips.
- Dwelling-related energy is higher per person in the city than in the lower density outer suburbs; however, the pattern for total household energy use is the reverse, with the suburbs accounting for higher household energy use.

The above result is explained, in part, by the higher proportion of single-person households in the inner city. In this context, the underlying drivers and policy responses require careful consideration. For example, if appropriate housing were available, it could facilitate single-person households sharing and, hence, reducing per capita energy consumption. Moreover, if a fixed number of single-person households are assumed at any one time, it may be more energy efficient if they are located near the inner city where the level of transport-related energy use is much lower. Also, reductions in residential energy consumption can be achieved with better housing design, use of improved appliances, and behavioral change. The complexity of the drivers and policy responses suggests the need for a much broader analysis incorporating sociological and cultural factors.

The model developed in this chapter will be useful for exploring several policy initiatives currently under consideration by the City of Sydney Council for a "2030 Vision of a Sustainable Sydney." This model will help explore the role that planning policy can play in achieving future targets, together with other initiatives being proposed, such as introducing Green Transformers[6] or smart meters (City of

Sydney 2009). The introduction of smart meters for water and energy usage will encourage consumers to reduce household energy consumption.

In general, this research has the potential to provide policy guidance to city officials in responding to climate change imperatives in carbon-constrained cities, including those in developing countries. As noted, the relatively simple data input requirements for the integrated urban model approach presented in this chapter would also allow it to be adapted to the context of developing countries, which are likely to have large data constraints.

The overall role of this model is to understand the citywide impacts of differences in energy consumption at the household level using household level data. This cross-scale analysis is unique and of vital importance in assessing how cities might respond to climate change imperatives. As part of a broader research analysis, the integrated model serves two important functions. First, it provides a spatial representation of greenhouse gas emissions across the city, which can be tracked through time to monitor climate change-related policy. Second, and more important, because of its ability to be configured for interactive and policy-relevant scenarios, the model lends itself to use as part of a deliberate process for improving the management and governance of cities. Such processes must involve government agencies, industry, and citizens in decision-making processes. By providing a single platform for water, energy, and transport data, the model can help overcome barriers of incompatible data formats between government data repositories.

In the context of infrastructure, built environment, and energy supply in cities, the model offers integrated answers to part of the question of resilience in the face of climate change. It supports policy-led approaches to efficient and effective planning, increasing the resilience and energy efficiency of carbon-constrained cities. The research also highlights the shortcomings of institutional and governance frameworks with respect to mitigation and adaptation priorities. Thus, the model has the potential to support the role of institutions and governance for improving the management, coordination, and planning of cities to meet climate change challenges.

Notes

1. http://www.whatiftechnologies.com
2. http://www.basix.nsw.gov.au/information/about.jsp
3. Depending on the level of information detail readily available, other demographic variables could be used in the construction of "household type."
4. Such a policy was recently enacted by the Australian government as part of its stimulus package in response to the global financial crisis, but this policy is not explored further in this chapter.
5. A more detailed explanation of the model structure, algorithms, and specifics of its inputs and outputs is presented in an earlier paper (Rickwood et al. 2007) and expanded in Rickwood 2009a (chapter 5) and Rickwood 2009b.
6. Green Transformers are cogeneration plants that convert waste to energy and produce low-carbon energy and recycled water as well.

References

ABS (Australian Bureau of Statistics). 2001. "Australian Bureau of Statistics 2001 Census of Population and Housing." http://www.abs.gov.au/websitedbs/D3310114.nsf/home/Census+data.

———. 2006. "Australian Bureau of Statistics 2006 Census of Population and Housing." http://www.abs.gov.au/websitedbs/D3310114.nsf/home/Census+data.

Baynes, T. 2009. "Complexity in Urban Development and Management: Historical Overview and Opportunities." *Journal of Industrial Ecology* 13 (2): 214–27.

Baynes, T. M., G. Turner, and J. West. 2005. *User Manual: Melbourne Region Stocks and Flow Framework*. Canberra, Australia: CSIRO Sustainable Ecosystems.

Beatty, R., S. O'Brien, and K. Beatty. 2006. "Fifteen Years of Drought, Demand Management and Pricing Reform in Urban Water. What's Gone Right, What's Gone Wrong and What's Needed for the Future?" Paper presented at Enviro 06 Conference and Exhibition, Melbourne, Australia, May 9–11.

City of Sydney. 2009. "Green Transformers." http://www.cityofsydney.nsw.gov.au/2030/thedirections/projects/EnviroPerfProj.asp, accessed April 5, 2009.

Echenique, M. H., D. Crowther, and W. Lindsay. 1969. "A Spatial Model for Urban Stock and Activity." *Regional Studies* 3: 281–312.

Echenique, M. H., A. D. J. Flowerdew, J. D. Hunt, T. R. Mayo, D. C. Simmonds, and I. J. Skidmore. 1990. "The MEPLAN Models of Bilbao, Leeds, and Dortmund." *Transport Reviews* 10: 309–22.

Hunt, J. D., D. S. Kriger, and E. J. Miller. 2005. "Current Operational Urban Land-Use-Transport Modeling Frameworks: A Review." *Transport Reviews* 25 (3): 329–76.

IPART (Independent Pricing and Regulatory Tribunal of New South Wales, Australia). 2004a. "Residential Water Use in Sydney, the Blue Mountains and Illawarra: Results from the 2003 Household Survey." Research Paper RP26, IPART, Sydney.

———. 2004b. "Residential Energy Use in Sydney, the Blue Mountains and Illawarra." NSW Independent Pricing and Regulatory Tribunal, Research Paper RP27, IPART, Sydney.

———. 2006. "Residential Energy Use in Sydney, the Blue Mountains and Illawarra." NSW Independent Pricing and Regulatory Tribunal, Research Paper RP28, IPART, Sydney.

Kenworthy, J. R., and F. B. Laube. 1999. *An International Source Book of Automobile Dependence in Cities, 1960–1990*. Boulder, CO: University Press of Colorado.

la Barra, T. D. 1989. *Integrated Land Use and Transport Modeling*. Cambridge, UK: Cambridge University Press.

Myors, P., R. O'Leary, and R. Helstroom. 2005. "Multi-unit Residential Energy and Peak Demand Study." *Energy News* 23 (4): 113–16.

Norman, J., H. MacLean, and C. Kennedy. 2006. "Comparing High and Low Residential Density: Life-Cycle Analysis of Energy-Use and Greenhouse Gas Emissions." *Journal of Urban Planning and Development* 132 (1): 10–21.

NSW (New South Wales) Department of Planning. 2005. "City of Cities: A Plan for Sydney's Future." NSW Government, Australia.

Putnam, S. H. 1983. *Integrated Urban Models*. London: Pion.

———. 1991. *Integrated Urban Models 2*. London: Pion.

Randolph, B., and P. Troy. 2007. "Energy Consumption and the Built Environment: A Social and Behavioural Analysis." City Futures Research Centre Research Paper 7, City Futures Research Centre, University of New South Wales, Australia.

Retamal, M., J. Glassmire, K. Abeysuriya, A. Turner, and S. White. 2009. "The Water Energy Nexus: Water and Energy Implications of Water Efficiency and Source Substitution Initiatives." Report by the Institute for Sustainable Futures Sydney to the Commonwealth Scientific and Industrial Research Organisation (CSIRO), Canberra, Australia.

Rickwood, P. 2009a. *The Impact of Physical Planning Policy on Household Energy Use and Greenhouse Emissions*. Unpublished Ph.D. thesis, University of Technology, Sydney, Australia.

———. 2009b. "Residential Operational Energy Use." *Urban Policy and Research* 27 (2): 137–55.

Rickwood, P., D. Giurco, G. J. Glazebrook, A. Kazaglis, L. E. Thomas, M. E. Zeibots, S. Boydell, S. White, G. Caprarelli, and J. McDougall. 2007. "Integrating Population, Land-Use, Transport, Water and Energy-Use Models to Improve the Sustainability of Urban Systems," State of Australian Cities National Conference (SOAC 2007), Adelaide, Australia. In *Proceedings of the State of Australian Cities National Conference 2007*, ed. S. Hamnett, 314–24. SOAC, Adelaide, Australia.

Roberts, P. 2005. "Yarra Valley Water: 2004 Residential End-Use Measurement Study." Yarra Valley Water, Melbourne, Australia.

TDC (Transport Data Centre). 2007. "2005 Household Travel Survey Summary Report—2007 Release." NSW Government Transport and Infrastructure, New South Wales, Australia.

Thomas, P. C., and L. Thomas. 2000. "A Study of an Energy Consumption Index Normalised for Area in House Energy Rating Schemes." In *Solar 2000, Proceedings of the Annual Conference of the Australian and New Zealand Solar Energy Society*, ed. D. Mills, J. Bell, L. A. Stoynov, and P. Yarlagadda, 113–21. Brisbane, Australia: Australian and New Zealand Solar Energy Society.

Thomson, J. M. 1977. *Great Cities and Their Traffic*. London: Gollancz.

Timmermans, H. 2003. "The Saga of Integrated Land Use-Transport Modeling: How Many More Dreams Before We Wake Up?" 10th International Conference on Travel Behaviour Research, Lucerne, Switzerland, August 10–15.

Troy, P., and B. Randolph. 2006. "Water Consumption and the Built Environment: A Social and Behavioural Analysis." City Futures Research Paper 5, City Futures Research Centre, University of New South Wales, Sydney, Australia.

Troy, P., D. Holloway, and B. Randolph. 2005. "Water-Use and the Built Environment: Patterns of Water Consumption in Sydney." City Futures Research Paper 1, City Futures Research Centre, University of New South Wales, Sydney, Australia.

US EPA (U.S. Environmental Protection Agency). 2000. "Projecting Land-Use Change: A Summary of Models for Assessing the Effects of Community Growth and Change on Land-Use Patterns." EPA/600/R-00/098, U.S. Environmental Protection Agency, Office of Research and Development, Cincinnati, OH.

Waddell, P. 1998. "A Behavioral Simulation Model for Metropolitan Policy Analysis and Planning: Residential Location and Housing Market Components of UrbanSim." In Proceedings of the 8th World Conference on Transport Research, Antwerp, Belgium, July 12–16.

———. 2002. "UrbanSim: Modeling Urban Development for Land Use, Transportation and Environmental Planning." *Journal of the American Planning Association* 68 (3): 297–314.

Waddell, P., and A. Borning. 2004. "A Case Study in Digital Government: Developing and Applying UrbanSim, a System for Simulating Urban Land Use, Transportation, and Environmental Impacts." *Social Science Computer Review* 22 (1): 35–51.

Wegener, M. 1994. "Operational Urban Models: State of the Art." *Journal of the American Planning Association* 60: 17–29.

Willis, R., R. A. Stewart, K. Panuwatwanich, B. Capati, and D. Giurco. 2009. "Gold Coast Domestic Water End Use Study." *Water: Journal of Australian Water Association* 36 (6): 79–85.

CHAPTER 5

Green Star and NABERS: Learning from the Australian Experience with Green Building Rating Tools

Lily M. Mitchell

[I]t is now recognized that energy efficiency of buildings and their climate change impacts goes to the heart of sustainability in the urban environment. —Baines and Bowman 2008 at 263

Green building rating tools can assist in addressing the climate change issues facing cities today by encouraging the development of more energy-efficient and resource-efficient buildings.

In Australia, two rating tools are in common use: the design rating tool Green Star and the performance rating tool NABERS (National Australian Built Environment Rating System). The two tools are very different in what they measure, yet their ratings (both expressed in number of "stars") are often confused. Calls have been increasing, including those from government, for convergence or at least standardization of the tools. Full convergence of these tools may not be achievable without sacrificing some valuable features of each tool, but there is potential for increased standardization, particularly in the metrics for energy use and greenhouse emissions. Internationally, too, there has been movement toward standardization of design rating tools.

Green rating tools will be best able to contribute to favorable environment and climate outcomes when they are in wide use, when they encourage building developers and operators to aim for ever higher performance, when they allow building users to easily compare buildings on environmental features and performance, and when they form part of a family of measures working together. The movement toward standardization or convergence of tools will mean that over the next few years a new generation of rating tools will be developed that will better realize their potential to achieve "green" building.

Lily M. Mitchell is a Senior Associate at Hanna and Morton LLP.

Introduction

The building sector is a significant source of greenhouse gas (GHG) emissions in Australia, as in other countries, with an energy demand responsible for almost a quarter of Australia's total emissions (CIE 2007). However, there is significant potential for cost-effective emission reductions in the building sector (Ürge-Vorsatz and Metz 2009; Warren Centre 2009; McKinsey and Company 2008; CIE 2007)—but there are barriers to be overcome before this potential can be realized. Appropriately designed green building rating tools can help overcome these barriers and can be an effective means of promoting energy efficiency and reducing emissions in the building sector.

Green rating tools use various methods to assess the potential, or performance, of a building in relation to specific sustainability criteria, usually including energy use as one of the central criteria. A building with a high green rating may use less than 70 percent of the energy used by an "average" building (Baines and Bowman 2008; ASBEC 2008)—a significant saving in emissions (and cost) over the life of the building.

A building that has obtained a green rating can then advertise its rating to tenants or purchasers interested in sustainable buildings. This allows developers of sustainable buildings to capitalize on their investment (Nelson 2008) and increases the awareness of building performance in the property market—and hence the demand for high-performing buildings (Cole et al. 2005; Campbell and Hood 2006; Hes 2007). Green rating tools have an advantage over mandatory building standards (though standards remain important) in that they give building developers and owners an incentive to build more energy- and resource-efficient buildings.

Demand for green-rated buildings is growing exponentially (Nelson 2008) and tools are rapidly being developed and disseminated—leading to competition between the tools, and confusion about the choices among tools. Another important development is that, although these tools were usually designed for voluntary use, governments and other agencies are increasingly requiring certain ratings for the buildings they occupy.

The effectiveness of these tools in the context of cities and climate change depends on several factors, including the way in which the tool is constructed, what it measures and how, the extent of acceptance and understanding of the tool in the building sector, the time and cost needed to obtain a rating, and the drivers (for example, government standards and community/tenant expectations) to obtain a high rating (see, for example, the criteria set out by Hes 2007).

The two Australian green building rating tools discussed in this chapter are Green Star, a design rating tool similar to tools used in the United Kingdom, the United States, and elsewhere, and the National Australian Built Environment Rating System (NABERS), unique to Australia, which benchmarks actual building performance. These tools differ on many of the above points. They have quite different approaches and objectives, yet their use in the relatively small Australian market has caused some confusion.

The experience of Australia with two different rating tools in common use may be of interest to others looking at ways to encourage efficiency and sustainability in buildings—particularly in light of the growing momentum, both within Australia and globally, toward standardization of green building rating tools, and the increasing use of such tools in mandatory measures.

Green Building Rating Tools Around the World

Green building rating tools are commonly used to assess and market new or refurbished buildings (primarily office buildings) in many developed countries. Table 5.1 lists some of these tools.

It is evident that many national tools share similarities with the world's first green building rating tool, BREEAM. Although the BREEAM, HK-BEAM, LEED™, Green Globes, and Green Star tools have a common ancestry and some similar features, the Comprehensive Assessment System for Building Environmental Efficiency (CASBEE) is somewhat different in approach, and NABERS and Energy Star are very different because they are performance ratings rather than design ratings. In many cases, these tools were developed by, or at the initiative of, the Green Building Councils in the relevant countries, members of the World Green Building Council. However, performance ratings (such as NABERS and Energy Star) are more commonly developed by government agencies.

Commencing with office building ratings, which remain the most commonly rated building type, the tools have been expanded to include rating systems for other

TABLE 5.1

Green Building Rating Tools Around the World

Jurisdiction	*Tool name*	*Tool abbreviation*
United Kingdom	Building Research Establishment Environmental Assessment Method (developed in 1990 and generally accepted as the world's first green building rating system)	BREEAM
United States and Canada	Leadership in Energy and Design	LEED™
	Green Globes	n.a.
	Energy Star (an energy performance benchmarking tool)	n.a.
Japan	Comprehensive Assessment System for Building Environmental Efficiency	CASBEE
Singapore	Green Mark	n.a.
Hong Kong, China	Building Environmental Assessment Method	HK-BEAM
Australia	Green Star	n.a.
	National Australian Built Environment Rating System	NABERS
New Zealand	Green Star	n.a.
South Africa	Green Star	n.a.

Sources: http://www.worldgbc.org/green-building-councils/green-building-rating-tools; http://www.greenglobes.com/; http://www.energystar.gov/; http://www.bca.gov.sg/GreenMark/green_mark_buildings.html; http://www.hk-beam.org.hk/general/home.php; http://www.nabers.com.au/; each last accessed February 7, 2010.
Note: Gray shading indicates that the tool is based on or similar to BREEAM. n.a. = not applicable.

classes of buildings, such as retail and education. In addition, separate tools are sometimes developed to rate different parts of a building at different stages of its life. For example, a tenancy may be able to be rated independently of the building as a whole, and a project may be able to be rated at the design stage or after completion.

Although much of the structure and methodology of a green building rating tool can be borrowed from a tool developed in another country, each country using such a tool has found it necessary to tailor the tool to the particular circumstances of the country—its minimum building standards, climate, and particular environmental concerns (for example, water use) (Saunders 2008).

Several commentators (see, for example, Hes 2007and Larsson 2004) have noted the different characteristics that can be used to describe building rating tools, which often indicate the different purposes for which they were developed and are used. These characteristics include whether a tool measures potential performance or actual performance, whether it is broad or focused in scope, and whether it is mandatory or voluntary. NABERS and Green Star can be seen to differ, to a greater or lesser extent, on each of these characteristics.

Building Code of Australia—Setting Minimum Requirements

Before discussing Australian green building rating tools in detail, it is worth noting that these tools are not the only method used to achieve resource efficiency and effectiveness in Australian buildings. Importantly, the Building Code of Australia sets minimum technical requirements for all building types across many areas of design and construction, including energy efficiency. The Building Code is maintained by the Australian Building Codes Board, an intergovernmental initiative with representatives from the building industry.

The Building Code is performance based: rather than specifying what materials or methods must be used, the Building Code sets out performance requirements for building materials, components, design factors, and construction methods, in order to attain certain broader objectives and functional statements. The performance requirements are mandatory, and compliance must be assessed by an authorized entity.

The Building Code is given effect by laws in each state and territory of Australia. In addition to incorporating the Building Code, these laws contain provisions on the submission of building plans, plan approval procedures, the issue of building permits, the accreditation of certain building materials, inspections during and after construction, the issue of compliance certificates, and the review and enforcement of standards.

Recently, Australian governments have requested that the Building Code be modified to increase the energy performance requirements for all new buildings from 2010. This will raise the baseline for new buildings, but there remains considerable room for differentiation between buildings and encouragement of greater improvements above the minimum standards. This is where green building rating tools play an important role.

Green Star (Australia)

Green Star is a rating tool designed for voluntary use to assess several environmental factors relating to a building's design. If the relevant criteria are met, a "Green Star rating" for that building may be awarded.

Key Features of Green Star

Table 5.2 provides a summary of the key features of Green Star.

How a Green Star Rating Is Calculated

In each of the categories, other than Innovation, a percentage rating of "points achieved" as against "points available" is calculated to give a category score. Then a state-specific weighting is applied to each category score to reflect the relative importance of that category. For example, in New South Wales (NSW), Energy and Indoor Environment Quality receive the highest weightings, at 25 percent and 20 percent, respectively. The sum of the weightings is 100 percent across the eight categories (GBCA 2009a). The weighting differs among states to reflect the differing importance of some issues across Australia, a large and climatically diverse country. For example, in the low-rainfall states, water conservation initiatives would receive a higher weighting.

After the category score is multiplied by the weighting for the relevant state, the weighted scores are totaled and any Innovation credits (which are not weighted) are added, giving a total score. As the Innovation credits are added to the existing scores, which themselves can total 100, a score of more than 100 is theoretically possible—but virtually impossible in practice.

Star ratings are awarded to certain ranges of point scores, as set out in table 5.3.

Independent Assessment and Certification

A Green Star technical assessment manual is prepared for each tool mentioned in table 5.2, describing each available credit and the compliance requirements to achieve the credit and providing further guidance and background information. These manuals are available for purchase from the Green Building Council of Australia (GBCA), generally at a cost of AU$500 for GBCA members and AU$600 for nonmembers. Any person with access to sufficient information about a building may use

TABLE 5.2

Green Star Features

Feature	*Description—Green Star*
Type of tool	Design rating
Development	First developed by Sinclair Knight Merz and the Building Research Establishment in 2003 (Saunders 2008, 27), the tool was then taken up and further adapted by the nonprofit Green Building Council of Australia (GBCA), a member of the World Green Building Council.
Supervising entity	Green Building Council of Australia
Basis for tool	Based on BREEAM as well as drawing on operational elements of the LEED system. Green Star, however, has been tailored to Australian conditions, such as climatic conditions and local building standards and regulations.
Purpose of tool	Assisting in differentiating and marketing buildings with strong environmental credentials—rather than as a tool to apply to every building. Intended for buildings in the top 25% of the Australian market (GBCA 2009a). As part of this approach, certified Green Star ratings are only issued if four or more stars have been earned. Green Star is largely a design rating, assessing the potential environmental performance of a new or refurbished building, or its attributes, rather than its actual performance or operation. This purpose is emphasized by the requirement that ratings be awarded within 24 months of completion of the building or the refurbishment. Green Star "assesses a developer's achievement without interfering operational factors such as building management and occupation" (Parker 2008, 2). These characteristics of Green Star are similar to those of the BREEAM, LEED, and Green Globes tools.
Available ratings	One to six stars—however, a certified rating will be issued only for ratings of four or more stars. No half stars are awarded. Results across the full suite of environmental categories are fed into one rating.
Costs to obtain a certified rating	The assessment fee charged by the GBCA varies between AU$5,200 for a small site (less than 2,000 m^2) where the applicant is a member of the GBCA, to AU$30,600 for a large site (greater than 100,000 m^2) where the applicant is not a member. In addition, the applicant may incur considerable internal/consultant costs in gathering information and preparing the application.

Types of tools currently available (by property type)	• *Office Design*—the most widely used tool in the portfolio, this tool assesses the environmental attributes of designs for office buildings in Australia. • *Office As Built*—similar to Office Design, but intended to assess the environmental attributes of newly built or refurbished office buildings in Australia, after completion of the project. • *Office Interiors*—designed to assess the environmental potential of an interior fitout. • *Retail Center*—assesses the environmental potential of new and refurbished retail centers in Australia. Ratings are assigned to the base building and its services, not to tenancy fitouts. • *Education*—assesses the environmental potential of new or refurbished education facilities in Australia. Base building and fitout are rated together (unlike other Green Star tools which assess base building and fitout separately). • *Multi-Unit Residential*—assesses the environmental potential of new or refurbished residential buildings containing two or more dwellings with over 80% of floor area for residential uses. • *Healthcare*—assesses the environmental potential of new or refurbished health and aged care facilities.
Development of new tools	Currently, the following tools are in the pilot phases of development: • Industrial • Mixed Use • Office Existing Building • Convention Centre Design Of these tools, Office Existing Building has caused some discussion, as it would closely overlap with performance ratings such as NABERS.
Eligibility criteria	A Green Star rating can be granted only if certain initial criteria are met. These include the following (GBCA 2009b): • *Spatial differentiation:* The project being rated is a distinct building and not a component of a wider project. • *Space use:* If a building has multiples uses (that is, multiple classifications under the Building Code), the building use being rated under Green Star, for example, education, office, or retail, must compose at least 80% of the gross floor area of the building. • Tool-specific *conditional requirements*, for example, that the building must not be sited on land of high ecological value or on prime agricultural land, or must not have greenhouse gas emissions greater than a specified level (for example, 110 kg carbon dioxide per square meter per annum, for Office Design version 3, the most recent version of the Office Design tool, estimated using modeling developed for NABERS Energy). • *Timing of certification:* As Green Star ratings relate to certain phases of a building's lifecycle, the building must be rated within a specified timeframe for particular ratings. For example, an application for an Office Design rating can be submitted before construction starts, whereas applications for As Built or Interior ratings can only be submitted after practical completion of the project. For both Design and As Built/Interior tools, ratings will be awarded only within 24 months of practical completion of the project (a project may be a new building or a refurbishment). These criteria are designed to ensure certain minimum standards and to further Green Star's aim of differentiating high-performing buildings. Therefore, they preclude Green Star ratings being used to compare designs across all types of buildings.

(continued)

TABLE 5.2

Green Star Features *(continued)*

Feature	*Description—Green Star*
Environmental categories assessed	If the eligibility criteria are met and the rating can proceed, points toward a rating can be scored in the following categories: • Energy • Transport • Water • Land Use And Ecology • Innovation • Indoor Environment Quality • Emissions • Materials • Management Each category contains a series of criteria that, if complied with, would reduce the environmental impact of a building in that category (Saunders 2008). For example, points in the Energy category can be earned for reducing emissions below the conditional requirement mentioned above, for submetering, for peak energy demand reduction, and for separate light switches for each zone, among other things (Green Star Office Design version 3). The full list is given in annex 5A. Innovation points are to encourage and recognize pioneering initiatives in sustainable strategies and technologies, for exceeding Green Star benchmarks, or for beneficial environmental design initiatives currently outside the scope of the Green Star rating tool. The inclusion of an "Innovation" category has been praised by commentators and is being adopted in other design tools such as BREEAM (Saunders 2008).
Revision of tools	New versions of Green Star tools are developed over time through a process of feedback, comparison against international standards, and revision—among other things—to raise standards in line with new conceptions of best practice.

Sources: GBCA 2009a, 2009b, http://www.gbca.org.au/, last accessed February 7, 2010.

TABLE 5.3
Green Star Ratings

Star rating	*Minimum assessment score*	*Comments*
One Star	10	Minimum practice
Two Stars	20	Average practice
Three Stars	30	Good practice
Four Stars	45	Best practice
Five Stars	60	Australian excellence
Six Stars	75	World leadership

Source: GBCA 2009a.

the manual to calculate the building's score under the Green Star system. However, to be able to publicize a Green Star rating and use the Green Star brand, the rating assessment must be certified. Certain advertisement rights are also given to project proponents who have registered with the GBCA for certification once their project has reached a stage when it can be certified.

Certification requires an independent panel, commissioned by the GBCA, to review the self-assessed rating and recommend (or oppose) the award of a particular Green Star certified rating. An assessment fee is also payable to the GBCA, as shown in table 5.2, which varies depending on the size of the project being rated and whether the applicant organization is a member of the GBCA. However, there may also be substantial expenditures associated with gathering the relevant information and preparing the application, which may cost AU$20,000 to AU$70,000 (Hes 2007).

Only ratings of four stars or more will be awarded a certification. Thus, while any building that fulfills the preconditions can assess its own Green Star rating, the Green Star brand can be used only in relation to relatively highly performing buildings. This restriction helps to fulfill one of the functions of the Green Star rating—a tool to market "green" buildings.

A sample certificate, awarded when a certified Green Star rating has been achieved, is included in annex 5B.

Prevalence of Green Star Ratings

The figures for certified and registered projects show that, while increasing proportions of new buildings are obtaining Green Star ratings, this growth is from a low base.

Data on the GBCA Web site (http://www.gbca.org.au/) indicate that, as of October 23, 2009, a total of 176 projects in Australia have received Green Star certified ratings; 134 of those—76 percent of the total number of ratings—were for Office Design, with the rest being distributed between Office As Built, Office Interiors, and pilot ratings for convention centers, education buildings, and multi-unit residential buildings. Figure 5.1 shows the breakdown by type of tool.

FIGURE 5.1
Breakdown of Green Star Ratings by Type, August 2009

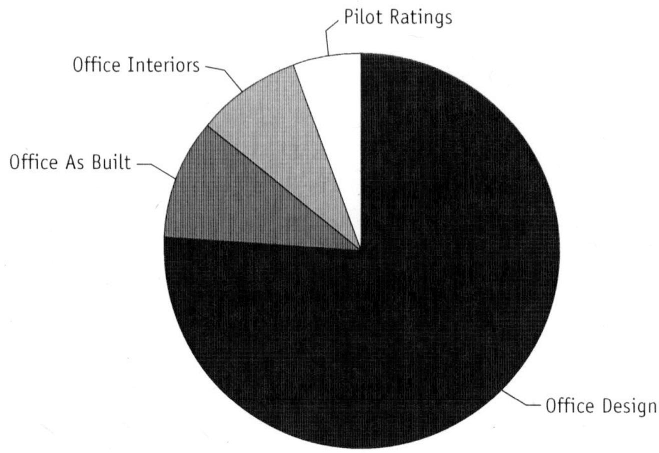

Source: http://www.gbca.org.au/greenstar-projects/, last accessed August 31, 2009.

These Green Star projects were concentrated in a few states—Victoria, Queensland, and NSW—which accounted for slightly over three-quarters of the total number of rated buildings, but there were Green Star–rated projects in every state and territory of Australia. Figure 5.2 shows the breakdown by location of project.

To indicate the spread of ratings, GBCA (2009a) provides a breakdown of the 125 ratings that were awarded up to January 2009 (including six for pilot projects): 46 of the ratings were for four stars, 62 were for five stars, and 17 were for six stars, as shown in figure 5.3.

Comparison to Other Design Tools

In tool design, Green Star is similar to BREEAM, as well as to LEED™—not surprising, since both Green Star and LEED were initially based on the BREEAM model. All three are predominantly design rating tools that aggregate points (credits) for specific design features across a wide range of environmental categories. While BREEAM was used as the basis for Green Star's methodology, the GBCA has since adapted Green Star's assessment methods such that they are now more similar to the LEED approach (Saunders 2008, 27).

In one study comparing building rating tools, Green Star was assessed to be comprehensive and rigorous (Campbell and Hood 2006). However, according to another comparative study, a building built to achieve the highest Green Star rating of six stars may, if located in the United Kingdom, only achieve a "good" or "very good" rating under BREEAM, rather than the top BREEAM ratings of "excellent"

FIGURE 5.2
Breakdown of Green Star Ratings by Project Location, August 2009

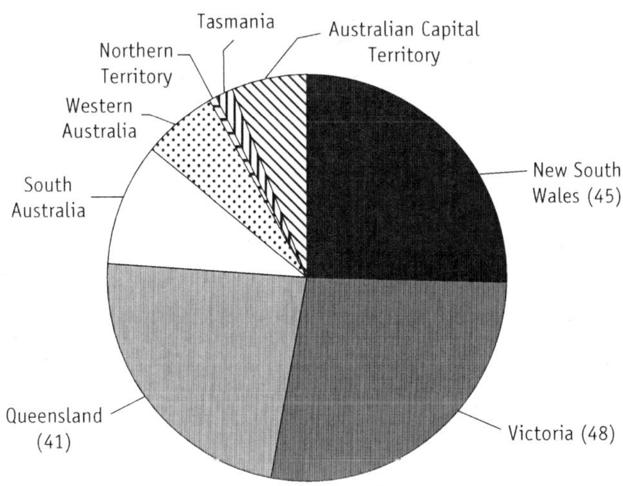

Source: www.gbca.org.au/greenstar-projects/, last accessed August 31, 2009.

FIGURE 5.3
Breakdown of Green Star Ratings by Star Rating, January 2009

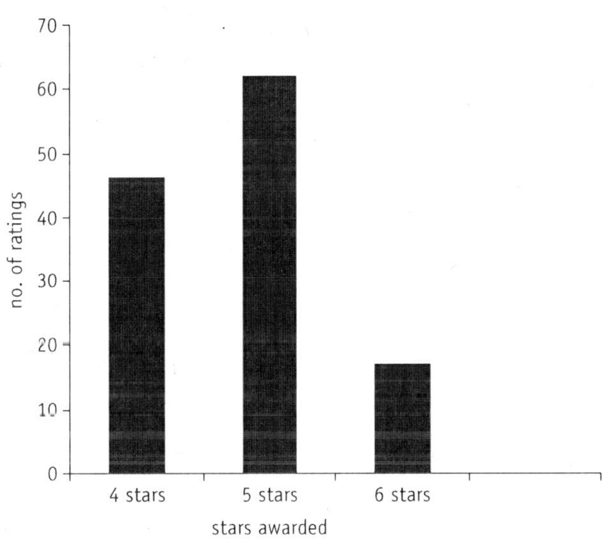

Source: GBCA 2009a.

or "outstanding" (Saunders 2008, 42). This may be due in part to differing minimum requirements set out in building codes, as Green Star awards points for design features that are better than requirements under the Building Code of Australia, but those requirements may be less stringent than their UK counterparts.

National Australian Built Environment Rating System

NABERS is a collection of separate tools, each of which calculates and rates the performance of an existing building (or part of one) on a particular environmental indicator at a certain point in time. Thus it differs crucially from Green Star, which rates design rather than performance. (On a simplistic level, the difference is that Green Star asks, among other things, "Does your building have separate light switches for each zone?", this being a design feature that can help reduce electricity use, whereas NABERS asks, "How much electricity did you use last year?")

Key Features of NABERS

NABERS has some similarities to the U.S. Environmental Protection Agency's Energy Star Portfolio Manager, but has broader application. The key features of NABERS are summarized in table 5.4.

How a NABERS Rating Is Calculated

NABERS ratings may be undertaken individually (for example, a company may decide to rate its building only in Energy and Water), and they are not combined into an overall rating, unlike Green Star. A further elaboration provided in the office rating tool is that a rating may be prepared for the whole building, the base building, or a particular tenancy. This provides useful flexibility, which is not available in many other rating tools.

Ratings for Energy and Water are calculated using data from 12 months of occupation/use of the building. For the NABERS Energy rating, the first step is converting the energy used by the relevant area in the 12-month period into GHG equivalents by reference to the emissions intensity of the standard energy mix used in the relevant State of Australia. (For example, if the building was located in Victoria, the emissions relating to its electricity use would be calculated with reference to the fact that in Victoria most electricity comes from high-emitting, brown-coal-fired power stations.) The raw emissions figures are then "normalized" to take into account the hours of use of the premises, the occupant and equipment density, and local climate. The normalized figures are then divided by the rated area, giving a figure expressing emissions per square meter. Finally, this figure is compared against the benchmark for the relevant state/territory and type of building (based on 10 years of data collection for this tool), resulting in a rating.

One interesting (and somewhat controversial) feature of NABERS Energy is that a higher rating of an existing building can be "purchased" if the building buys renewable energy (through the Green Power scheme). This option is not available for new buildings.

TABLE 5.4

NABERS Features

Feature	*Description—NABERS*
Type of tool	Performance rating
Development	NABERS incorporates the tool formerly known as the Australian Building Greenhouse Rating tool (ABGR), now known as NABERS Energy, for offices, developed in 1999 by the NSW Sustainable Energy Development Authority. ABGR became part of the NABERS suite of tools in 2008 and has since been expanded to include Water, Waste, and Indoor Environment ratings and to cover a range of building types.
Supervising entity	NSW Government, Department of Environment, Climate Change and Water (DECCW). Operations are overseen by the NABERS National Steering Committee, which comprises representatives of the Australian and State and Territory Governments, with the Australian Sustainable Built Environment Council as an observer.
Basis for tool	Unlike Green Star, ABGR/NABERS tools were developed from scratch.
Objectives	To provide a framework for improving the environmental performance of buildings. The emphasis is on relevance, realism, and practicality (http://www.nabers.com.au/page.aspx?cid=534&site=2, last accessed October 23, 2009).
Eligibility criteria	No criteria such as those required for Green Star ratings. However, to obtain a NABERS Energy or Water rating, 12 months of performance data on the relevant environmental category are required.
Available ratings	One to five stars, with half stars available. Separate rating for each environmental category (see below).
Costs to obtain a certified rating	Assessment fee charged by accredited assessor, averaging approximately AU$5,000–$6,000, depending on the area being rated, plus a lodgement fee of approximately AU$1000. These amounts are per environmental category, but discounts may apply if several environmental categories are being rated at the same premises.
Types of tools currently available (by property type)	• *Office buildings*—tenancy, base building, and whole building can each be separately assessed for each environmental category. • *Hotels*—assesses performance of hotels (including facilities, such as function rooms and swimming pools) on energy use and water use. • *Residential buildings*—assesses performance of individual dwellings on energy and water use. This rating does not currently apply to apartments, or other dwellings where services (such as pools, gyms, underground car parks) are shared or utilities are not separately metered.

(continued)

TABLE 5.4
NABERS Features (*continued*)

Feature	*Description—NABERS*
Development of tools for new property types	NABERS tools for the following types of properties are planned to be released in the future: • shopping centers • schools • hospitals • data centers
Environmental categories currently assessed	For each building type, the following environmental aspects can be rated separately (with details taken from the NABERS Office rating tool): • *Energy:* The amount of each type of energy (electricity, gas, coal, oil) consumed on the premises in a year, and how much of it is supplied from "Green Power" (renewable energy that can be voluntarily purchased from electricity retailers in respect of electricity and gas use). This is by far the most commonly used rating. • *Water:* The amount of water used on the premises in a year, and how much of this is externally supplied, recycled water. This tool is also becoming more widely used. • *Indoor Environment:* Requires subratings of the premises' thermal comfort, air quality, acoustic comfort, lighting, and office layout. • *Waste:* The total materials used (for example, paper) per person per day and the amount of those materials that are recycled or reused. A relatively new addition to the suite of tools. NABERS is one of the broadest performance rating tools available, though it does not cover as many environmental factors as design rating tools such as Green Star.
Future development	In addition to the above categories, a Transport category is currently being developed, and the NABERS Web site notes that other elements will be developed to enable buildings to be rated on a full range of measured operational impacts, including: • refrigerants (greenhouse and ozone depletion potential) • stormwater runoff and pollution • sewage • landscape diversity (http://www.nabers.com.au/page.aspx?code=ABOUTUS&site=2, last accessed October 23, 2009) Once completed, this suite of measures will cover similar topics to those assessed as part of a Green Star rating.
Revision of tools	NABERS tools are periodically reviewed by expert committees. Reviews consider both whether corrections are needed to ensure the benchmarks for each star level remain fair and comparable, and whether an additional star is required to represent a new standard of exceptional performance.

Source: http://www.nabers.com.au, last accessed October 23, 2009; DECCW 2009; McAteer 2008; Ostoja 2008, except where otherwise indicated.

NABERS Water follows a similar process to NABERS Energy, using water bills instead of energy bills. For Indoor Environment, metering of indoor environment conditions and an occupant satisfaction questionnaire are required. For Waste, a daily waste audit is conducted for 10 consecutive working days.

Ratings for each component are expressed in stars, as with Green Star, but the maximum number of NABERS stars is five (rather than the six in Green Star). Half-stars are available, allowing greater discrimination on performance than the whole stars used in Green Star.

Table 5.5 provides comments on what each whole star rating represents. The normalized emissions per square meter that would result in such a rating will differ depending on the state or territory in which the rated area is located and whether the rated area is the base building, a tenancy, or the whole building. As an illustration, table 5.5 lists the normalized emissions per square meter for a base building in NSW.

As it rates actual performance, which may vary over time, NABERS is a point-in-time rating tool and its ratings remain valid for only one year from the date of the rating.

The normalization and benchmarking process is reviewed periodically (for example, Ostoja 2008), and recommendations for corrections are made where necessary to ensure that the rating bands reflect current performance, with median performance earning 2.5 stars, that they allow for superior performance, and that they are fair and comparable. For example, the 2008 review recommended some changes to the system for adjusting for the different energy sources and climate of different states and territories of Australia, to allow greater comparability between buildings in different States (Ostoja 2008).

Self-Assessment or Accredited Assessment

The NABERS Web site (http://www.nabers.com.au) provides some calculation tools to allow entities to calculate the rating of any building for which they have 12 months of data—at no cost. Calculation of an office building's NABERS Energy rating, for example, can be a simple process of entering the address and size of the area to be rated, its operating hours per week, the number of people and computers on the

TABLE 5.5

NABERS Star Ratings (Energy)

Star rating	*Comments*	*Emissions (kg CO_2/m^2)*
One star	Poor—poor energy management or outdated systems	199
Two stars	Average building performance	167
Three stars	Very good—current market best practice	135
Four stars	Excellent—strong performance	103
Five stars	Exceptional—best building performance	71

Source: DEWHA 2008.

site, and the amount of electricity and gas used over 12 months (assuming the building does not use any other types of energy).

However, in order to use the NABERS trademark, a building must receive an official rating calculated by a NABERS-accredited assessor (a list of assessors is available on the NABERS Web site). The assessor, with the building owner's assistance, will collate the relevant data (original documents are required), enter the data into a spreadsheet, calculate the rating, and submit the rating to the NABERS national administrator for auditing and certification (McAteer 2008). The assessors charge fees as shown in table 5.4; these fees vary depending on the tool, the premises, and the environmental category being rated. A sample certificate, awarded when an accredited NABERS rating has been achieved, is provided in annex 5C.

Use of NABERS in Design Phase

Although NABERS is a performance rating tool, a property owner/developer can enter into a commitment to obtain a certain NABERS Energy rating at the design stage. To do so, the owner/developer enters into a commitment agreement with the Department of Environment, Climate Change and Water (DECCW) that specifies the building and the target star rating (which must be at least four stars). The owner/developer agrees to

- design and construct the building to operate at the specified NABERS level.
- inform all its consultants, contractors and tenants of this commitment.
- have an independent design review undertaken to determine whether any changes to the building's design are necessary in order to achieve the target rating.
- obtain a NABERS rating within 13–14 months from the date of full operation of the building.

After signing the agreement and paying a fee to the DECCW, the owner/developer can use the NABERS trademark to promote the "commitment rating" of the development.

Prevalence of NABERS Ratings

The NABERS Energy rating is widely used, particularly in relation to large office buildings. The DECCW estimates that 41 percent of the national available office space (approximately 8.6 million square meters) has obtained NABERS Energy ratings—more than 600 buildings (DECCW 2009).

Although there are rated buildings across Australia, NSW has a greater rated area than any other state or territory, which may possibly have to do with the fact that NABERS was first launched in NSW and is administered by a NSW body. Figure 5.4 provides the percentage of total net lettable area in each state/territory that has been rated with NABERS Energy. In Western Australia as well as NSW, half of the total net lettable area in the state has been rated.

FIGURE 5.4
Percentage of Total Net Lettable Area Rated with NABERS Energy, by State/Territory, June 2009

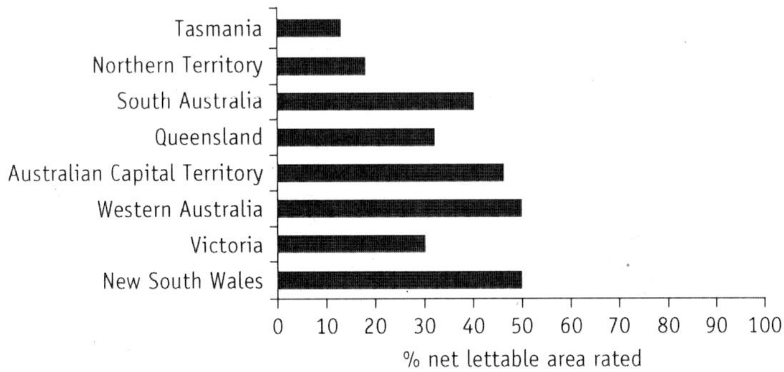

Source: DECCW 2009.

Ratings of tenancies are much less common than base building or whole building ratings (DECCW 2009).

Spread of NABERS Ratings

The NABERS benchmarks are calibrated such that a 2.5 star rating represents the median performance of Australian buildings.

The NABERS Web site provides information on office buildings with current accredited energy and water ratings. Figure 5.5 shows the distribution of accredited NABERS Energy star ratings for office buildings, with the most common rating being five stars (81 buildings). Rather than producing the classic bell curve, this graph indicates that higher performing buildings are more likely to seek ratings. This trend has increased in the past few years: of the NABERS Energy ratings submitted in 2008, 41 percent of the base building or whole building ratings, and more than 60 percent of the tenancy ratings, were four stars or higher (DECCW 2009).

Case Study: Energy and Greenhouse Benefits of NABERS Ratings

Considerable energy savings and consequent GHG emissions reductions can be demonstrated through the use of NABERS Energy ratings (table 5.6). DECCW (2009, 2) notes that, "Many office buildings use accredited NABERS Energy ratings to track ongoing greenhouse performance and improvement through annual ratings. Together, these buildings are saving a combined 149,000 tonnes of greenhouse gas emissions every year—an average 15 percent savings compared to their first rating."

FIGURE 5.5

Breakdown of Current NABERS Energy—Office Ratings by Star Rating, August 2009

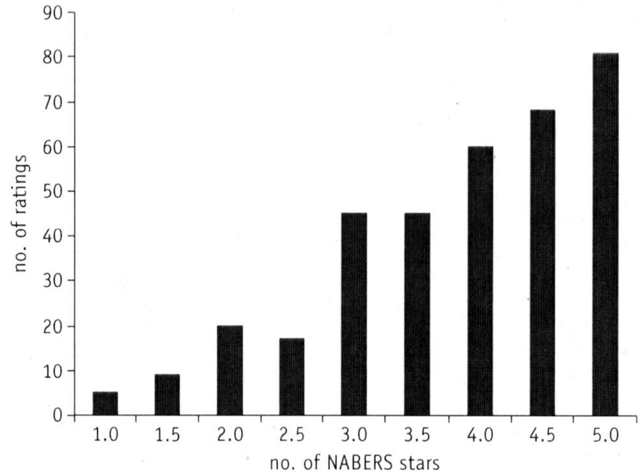

Source: http://www.nabers.com.au/frame.aspx?show=building&code=Buildings&site=2, last accessed 31 August 2009.

TABLE 5.6

Energy and Greenhouse Savings Demonstrated By Re-ratings of Buildings

Feature	*Figure*
Number of buildings	381
Total area (m^2)	6,288,000
Tonnes CO_2 saved per annum	149,000
Average NABERS Energy star improvement	0.3
Average % CO_2 savings	15%
kg CO_2/m^2 saved	24

Source: DECCW 2009.

Comparison to Other Performance Tools

Although various tools have been developed in different countries to measure building energy use, there are few full-fledged performance rating tools similar to NABERS. The U.S. Energy Star program may be most comparable. Energy Star, originally developed to rate the energy efficiency of computers, now has a Portfolio Manager tool that is used to rate buildings under the Environmental Protection Agency's National Energy Performance Rating (NEPR) system. Table 5.7 illustrates how some of the key features of Energy Star for buildings compare to those of NABERS.

TABLE 5.7

U.S. Energy Star Compared with NABERS for Buildings

Feature	*Description—Energy Star*
Supervising entity	As with NABERS (and unlike Green Star and LEED), Energy Star is run by a government agency, the U.S. Environmental Protection Agency, rather than a private enterprise.
Data requirements	Data required for a rating is similar to that required under NABERS, including 12 months of energy data and basic building information.
Type of rating	NEPRs are expressed as a rank between 1 and 100, with 50 indicating average performance and 80 or higher indicating performance in the top 20% (compared to a benchmark established through national surveys). Ratings are normalized for various factors such as weather conditions, number of building occupants, and hours of occupation—as is also the practice with NABERS.
	If a building receives an NEPR of 75 or higher and also meets industry standards for indoor air quality (as validated by a professional engineer), it can display the Energy Star label to promote its high performance. Unlike NABERS, the Energy Star label is not available in different levels to indicate differences in performance, but the label does include the NEPR, which allows finer distinctions.
Types of buildings that can be rated	Eleven different types of building can be rated under Energy Star (more than under NABERS), including schools, hospitals, banks, warehouses, and supermarkets. As with NABERS, office buildings remain the most commonly rated building type.
Environmental categories	While the focus of Energy Star is on energy, water use can also be tracked using the Portfolio Manager tool. NABERS offers more categories to assess, namely indoor environment and waste.
Costs of rating	Costs may be lower than those of NABERS. The only costs are those charged by the verifying engineer, which are estimated to be between US$0.005 to US$0.01 per gross square foot (Source: http://energystar.custhelp.com/cgi-bin/energystar.cfg/php/enduser/std_adp.php?p_faqid=2508&p_created=1147193775&p_sid=f00ij2Lj&p_accessibility=0&p_redirect=&p_lva=&p_sp=cF9zcmNoPTEmcF9zb3J0X2J5PSZwX2dyaWRzb3J0PSZwX3Jvd19jbnQ9MzAsMzAmcF9wcm9kcz0yOTkmcF9jYXFzPSZwX3B2PTEuMjk5JnBfY3Y9JnBfcGFnZT0x&p_li=&p_topview=1, last accessed October 23, 2009).

Source: http://www.energystar.gov/, last accessed February 7, 2010, except where otherwise indicated.

Key Differences Between Green Star and NABERS

NABERS and Green Star are very different in purpose and application. Table 5.8 summarizes key differences, some of which are then discussed in more detail below.

Relative Costs and Ease of Use of the Tools

Given the differences between the NABERS and Green Star tools and their different uses, it is not surprising that opinions on the values and virtues of these ratings tools vary.

Although Green Star is based on the internationally accepted and widely used BREEAM and LEED tools, there is a perception that it is impractical in some circumstances. It is time consuming to amass the supporting information required to substantiate sufficient points to earn a high Green Star rating. This is the case particularly in relation to As-Built tools—Design ratings are easier in this respect, and this may be one reason why so many more Green Star Design ratings have been awarded

TABLE 5.8
Key Differences Between Green Star and NABERS

Feature	*Green Star*	*NABERS*
Type of tool	Design—rates potential of building	Performance
Supervising entity	Private entity—the GBCA	Government entity—the DECCW
Aim	To distinguish buildings with good environmental design features; only intended for top 25% of market (GBCA 2009a)	To identify, and enable comparisons of, building performance
Factors assessed	Broad—energy, transport, water, IEQ, materials, land use, management, innovation—all assessed for one rating	Specific—energy/water/waste/ IEQ—each rated separately
Assessment method	Accrue points for specific features in each category	Performance data compared to benchmark
Main application	Design phase	Operational phase
When rating can be obtained	In design phase or within 2 years of completion of construction/ refurbishment	When building is completed and occupied—usually 12 months of operational data are required
Duration of rating	Does not expire	Expires 12 months after date of rating
Cost of accredited rating	Fees AU$5,200–AU$30,600 (depending on size of rated area), plus costs of preparing application	Fees approximately AU$5,000–6,000 for average size premises, plus lodgement fee of approximately AU$1,000

Sources: GBCA 2009a, 2009b, http://www.gbca.org.au/, last accessed February 7, 2010; www.nabers.com.au, last accessed October 23, 2009; DECCW 2009; McAteer 2008; Ostoja 2008.

than any other kind. It is perceived to be difficult to obtain a high rating without incurring increased building costs and reducing the flexibility of use of the building. Actions to obtain some Green Star points, such as reducing the number of car spaces or reducing nighttime lighting, may be seen as reducing the value of the property, while others depend on external factors and cannot be achieved by design alone (such as being close to a public transport node). These comments, particularly in relation to the time and cost of obtaining a rating, echo those expressed in studies of other similar rating tools, such as LEED and BREEAM (see, for example, Larsson 2004; Smith et al. 2006; Hes 2007). It is possible to develop cheaper, easier-to-use online versions of design rating tools, as Green Globes has aimed to do in the United States. Green Star does have an online self-assessment option, but it does not lead to a certified Green Star rating. The cost of obtaining a Green Star rating is likely to mean that it will be used only, or predominantly, in developed countries, and then only at the top of the property market.

A NABERS rating can be quicker and cheaper to obtain (once the requisite 12 months of data are available) and does not have the preconditions and restrictions of a Green Star design rating. A noncertified NABERS rating can also be calculated online, at no cost. Therefore, NABERS may be better suited to a broad-based push to rate and compare the performance of buildings. However, to obtain annual NABERS ratings across the full suite of indicators (so as to be more comparable to Green Star as a broad sustainability measure) would still require some investment of time.

Design or Performance Rating?

Design rating tools and performance rating tools each have their strengths and weaknesses, and it is important to note that what one participant in the building sector sees as a strength at one phase of a building's lifecycle may be a weakness to another participant or in another phase.

As a design rating, Green Star encourages sustainable decision making at the design stage, which is crucial for the overall sustainability of the building. However, it does not provide any incentive for efficient management once the building is in use. Performance in practice may not be as good as the potential, particularly in relation to ongoing energy use where building management and tenant activity play an important role (Hes 2007).

NABERS focuses attention on the commissioning, operation, and maintenance of a building, which are key factors in ensuring efficiency over the long operating lives of buildings (DEWHA 2008). It also captures the impact of decisions made during the design and construction phases, and thus complements design ratings (McAteer 2008). As a NABERS rating is only valid for one year, a building may obtain many NABERS ratings over its life and is rewarded with a higher rating if its performance improves over time.

In terms of educational aspects, in Green Star the detailed process of accruing points in each category toward a rating helps building designers and developers learn about sustainability features, and makes it clear where there are areas in which more points could be earned. The simpler NABERS process does not indicate, in and of itself, to building operators where or how improvements can be made.

While a design rating might be thought most appropriate when advertising a building as yet unbuilt or being built, both NABERS and Green Star allow for registration of a development as aiming to achieve a rating, prior to the time at which an official rating can be obtained, thus allowing for early marketing of the building as "green." (When a building is able to be rated, it must obtain a rating or cease using the symbols of the relevant rating tool.)

The appropriateness of a rating tool for use with existing buildings (rather than just new or significantly renovated buildings) is now receiving increased attention, given the importance of upgrading existing building stock (Campbell and Hood 2006; DEWHA 2008). A performance rating tool such as NABERS is well adapted for this purpose. Green Star ratings are currently restricted in this regard, as they can be awarded only within two years of completion of the building (or the renovation project). However, a Green Star Office Existing Building rating tool is now being developed to address this issue.

One virtue of the Green Star tool that is not shared by the NABERS tools is the ability to take into account wider factors (including some relevant to supply chain and life cycle assessment) such as the sustainability of the materials used in building construction, the treatment of construction waste, the siting of a building, and transport links to the building. (Larsson 2004 notes that transport emissions in the journey to and from a building are of the same order of magnitude as the building's operating energy.) NABERS currently cannot assess these factors. It may be possible to build performance ratings for some of these issues, but collecting actual performance data is likely to be difficult and time consuming.

Impacts of Building Ratings on Building Value

Improved environmental ratings of Australian buildings have not yet been reliably shown to increase the rents paid for those buildings. A study by property management company Jones Lang LaSalle (2006, p. 6) found that, "whilst tenants currently may not be willing to pay a premium rental for buildings with sustainability features, some tenants will very soon come to expect a discount to occupy buildings that do not have these features." More recently, a GBCA report (2008, p. 5.) noted that, "While some tenants are willing to pay the rental cost of achieving Green Star, a rental premium is not yet proven in all cases."

However, a variety of other positive impacts have been observed; these are summarized in table 5.9.

Moving Beyond Voluntary Use of Green Rating Tools

In addition to voluntary use of the Green Star or NABERS rating tools by private parties seeking to market their building as environmentally friendly, it is increasingly the case that certain ratings are required by various agencies. As noted by Larsson

TABLE 5.9
Benefits Associated with High Green Building Ratings

Green Star	*NABERS*
• Approximately two-thirds of interviewed property investors said they would pay more for a Green Star building (p. 17). • "Long-term rental growth, tenant retention and operating cost savings are the key drivers of the increasing market value of green buildings" (p. 18). • "The improved marketability of Green Star buildings is their main current competitive advantage: they are easier to sell and lease, which reduces vacancy times and hence income losses" (p. 5). • "Green Star buildings have achieved a reduced capitalization rate to the order of 0.25-0.50% when compared with the rest of the market" (p. 20).	Sample building: a 30,000 square meter A-grade office building in central Sydney, with a single tenant. Undertaking relatively simple improvements to building services that would raise the building's NABERS Energy rating by one star would result in • an increase in the building's capital value of approximately AU$3 million—a return of almost 10 times the required investment. • a decrease in outgoings by AU$3.32 per m^2, leading to $99,700 savings in annual outgoings. • a reduced letting period.

Sources: For Green Star, GBCA 2008; for NABERS, JLL 2006.

(2004) and Hes (2007), rating tools with a narrow focus and more objective measurements are more likely to become part of regulatory systems than the broader, more aspirational rating tools. In Australia, both NABERS and Green Star ratings are now being required in certain circumstances, but NABERS has been used more widely, as noted in the section "Prevalence of NABERS Ratings," perhaps partly because it fits the narrow focus and more objective measurement criteria Larsson and Hes noted.

Concerns have been expressed about the tendency for hitherto voluntary standards to become part of mandatory regulation, particularly where the relevant tools are developed and administered by commercial, nongovernmental organizations (such as the GBCA, whose paying members include the organizations required to have their properties rated). Accountability, transparency, and industry capture issues may arise (Baines and Bowman 2008), though there is no suggestion that this is currently an issue with Green Star and the GBCA. NABERS may be less of a concern in this regard, as it is government operated.

Current Requirements to Hold Ratings

Table 5.10 summarizes current requirements for various entities and organizations to hold specific ratings.

Upcoming Mandatory NABERS Energy Disclosure

In an important development, the Australian government has proposed to require the owner/lessor of an office building to obtain a NABERS Energy rating (with

TABLE 5.10

Rating Requirements for Organizations

Relevant organization	*Requirement*
Australian governments	The Australian Government and several state/territory governments have introduced NABERS Energy requirements for office premises owned or leased by government. This is an important market driver, given that the Australian government alone represents approximately 13% of the commercial office market (COAG 2009), and taken together the nine Australian governments occupy approximately 25% of the commercial office market. Under the Energy Efficiency in Government Operations Policy of the Australian Greenhouse Office, Department of the Environment and Water Resources, Commonwealth of Australia (2006), the Australian government requires a 4.5 star NABERS Energy rating for new buildings and major refurbishments, and all new leases must include a requirement for annual NABERS Energy ratings. The policy notes (p. 16) that NABERS Energy was adopted "as the preferred rating tool due to its broad acceptance by the industry and access to a low cost independent performance certification scheme." State/territory government rating requirements range between 3 and 5 stars, depending on the type, size, and age of the property (see, for example, the NSW Government Sustainability Policy, available at http://www.environment.nsw.gov.au/resources/government/08453 SustainabilityPolicy.pdf; Victorian Government's Sustainability Action Statement 2006, at 80, each last accessed February 7, 2010). However, these standards are not always complied with in practice. There is also discussion of moving toward a consistent green leasing policy to apply to all government tenancies in office buildings over a certain size across Australia, with a prescribed minimum NABERS Energy rating (for example, 4.5 stars, as per the Australian Government standards). However, this policy may take some time to take full effect, as it is proposed to apply only to new and renewed leases.
Property Council of Australia (PCA)	NABERS and Green Star ratings are included as criteria in the PCA's "Guide to Office Building Quality" (2006). For a building to be considered Premium or Grade A under the PCA criteria (top-quality buildings that attract significantly higher rentals than lower grade buildings), it must have, among other criteria, a NABERS Energy rating of at least 4.5 stars and a Green Star rating of at least 4 stars. For Grade B, 4 NABERS Energy stars and 3 Green Stars are required.
City of Sydney Council	The "Draft Ecological Sustainable Development—Development Control Plan 2007" requires new or refurbished office buildings with a net lettable area greater than 1,000 m^2 to have a minimum of 4.5 stars under NABERS Energy, and new office buildings are also required to have a minimum of 4 stars under the Green Star Office Design and Office As Built tools (City of Sydney Council 2007).
NSW Energy Savings Scheme (NESS)	NESS employs NABERS Energy for office buildings to normalize the energy use baseline for a demand-side abatement project in an office building (Energy Savings Scheme Rule 8.8, available at http://www.ess.nsw.gov.au/documents/ESSRule.pdf, last accessed February 7, 2010).

Sources: Cited for each entry.

certain modifications) on the sale or lease of an area of 2,000 square meters or more, from 2010 (DEWHA 2008). The rating, in the form of a certificate, must be disclosed to potential purchasers/tenants of the premises. This is similar to the requirement in Europe to disclose Energy Performance Certificates on sale or lease, under the Energy Performance for Buildings Directive (2002/91/EC). The draft form of the certificate that is to be disclosed is shown in annex 5D.

While NABERS Energy is currently the only rating tool proposed for use under this scheme, DEWHA (2008) notes that other tools may be considered at a later stage. Green Star (at least in its current form) was not considered appropriate, as it assesses predicted rather than actual performance, and thus does not allow comparison of the actual energy efficiency of buildings. It also includes a wider range of considerations than were thought relevant for the mandatory disclosure scheme (DEWHA 2008).

A requirement that building owners must obtain and disclose a green building rating can assist in overcoming the principal-agent and information market failures identified as some of the reasons why building energy efficiency measures are so underutilized (DEWHA 2008; ASBEC 2008). It raises awareness among building sector customers of the differing performance of different buildings, making it easier for them to compare buildings and take into account a building's "green" performance in deciding whether to buy or lease a property—and what price to pay for it.

Requiring these ratings may also influence the landlord to increase the energy efficiency of the building (Nelson 2008). In fact, a recent study from a large sample of buildings found that buildings that disclose their NABERS Energy ratings to their tenants perform better, to the extent of half a NABERS Energy star, than those that do not (Warren Centre 2009).

Can the Rating Tools Be Integrated?

Green Star and NABERS have different aims and approaches. Green Star is intended to distinguish new or refurbished buildings with the potential for above-market environmental performance across a full range of indicators. NABERS tools offer a snapshot of how a building, or part of one, is performing on specified environmental indicators (for example, energy use or indoor air quality) at a point in time, allowing comparability between premises.

However, Evans and Wotton (2008) state that, although the trademarks of both Green Star and NABERS are well recognized and accepted, the existence of two schemes "has caused some confusion among building owners and tenants." This has been echoed in comments made by representatives of the GBCA and the PCA (Lenaghan 2008). The fact that both tools use "stars," yet the stars are awarded for different attributes, and an equal star rating on both tools does not mean equal performance, tends to increase confusion, in part because it is not always clear which type of star rating is being referenced.

A Possible Solution to User Confusion

Property market commentators and stakeholders in the rating tools are discussing whether and how to integrate the two tools in order to address the confusion in the marketplace. They recognize that if the issue of having two very different but competing rating tools can be addressed, the integrated design/performance tool would be attractive for use around the world, as no other country has yet resolved this issue (Lenaghan 2008).

The Council of Australian Governments (COAG) recently announced a plan to develop a "consistent outcomes-based national building energy standard setting, assessment and rating framework," to be implemented in 2011. This framework is intended to apply to all types of buildings, commercial and residential, new and existing, and will "work towards convergence of existing, measurement-based rating tools (such as the National Australian Built Environment Rating System—NABERS Energy) for existing buildings with predictive or modeling-based tools used for rating new buildings," as well as setting minimum energy performance standards (COAG 2009, 23).

This is the strongest statement yet that convergence or integration of the rating tools, at least in the area of energy, is to be expected. But how this convergence will be achieved must be addressed.

How the Rating Tools Can Be Integrated

It is worth bearing in mind that there is a range of actors in the building sector, and they will need different things from a rating tool depending on their role, the type of building, and the phase of the building's lifecycle, among other elements (Campbell and Hood 2006). This is an argument for retaining the most useful aspects of both design and performance rating tools in order to be able to choose a tool appropriate in the circumstances.

The Australian government, or other entities engaged in developing or standardizing rating tools, should consider ways to reduce confusion between the rating tools, to enable users to pick the tool best suited to their purpose, and to retain the different valuable features of each tool while allowing the greatest possible comparability. This does not necessarily require complete convergence of the tools. What other steps could be taken?

It will not be possible to merge design and performance rating tools completely, given their different natures and purposes. Recognizing this limitation may help to reduce market confusion by making the ratings more distinct, so that a Green Star rating is not confused for a NABERS rating. A change of terminology might assist—for example, the high performance of a Green Star–rated building across a wide suite of issues might be better conveyed with labels such as Silver/Gold/Platinum, as used by LEED, whereas the more specific assessments of NABERS tools for individual issues (which are not limited to top performers but can report the full gamut of performance) is better suited to retain some form of numerical labeling.

The fact that Green Star is an all-encompassing rating whereas NABERS has separate ratings for separate issues could be addressed, if thought desirable, by amending either or both of the tools as follows:

- Amending Green Star so that it provides subratings for each issue, based on the category score.
- Amending NABERS so that it provides overall ratings based on some combination of the individual issue ratings.

The first option may well be adopted, in relation to Green Star's energy category, as part of COAG's proposed new energy framework. The second option would only be possible if owners of the relevant building decided to obtain NABERS ratings for each issue—which is not commonly done.

One suggestion to assist in comparability is that a design rating of a building should be followed, at the appropriate time, with a performance rating using the same assessment structure to determine the extent to which the potential of the building has been fulfilled in practice (Larsson 2004; Campbell and Hood 2006). This is not always possible, given that design and performance ratings necessarily use different indicators and are sometimes measuring different things.

However, where practicable, measurement criteria should be standardized. There is at least one instance in which Green Star and NABERS already coincide: Green Star energy use modeling (at the building design stage) includes an option to use NABERS Energy metrics. No doubt more areas for standardization could be identified—perhaps including the treatment of differing energy sources and climatic conditions in different states/territories. But there is much work to be done—and some entrenched positions on behalf of the designers and administrators of the different tools to be reconciled—before a significant degree of standardization can be implemented.

Single Responsible Agency

The standardization process may move more swiftly and effectively if one agency were to be responsible for both types of rating tools. It may also assist in educating users of the tools (both those trying to choose a rating tool to use for their buildings and those trying to understand existing ratings) if all information relating to the tools were available in one place. Having a single agency in charge would make it easier for the regular reviews and updates of the tools to be done concurrently and would allow for easier cross-referring. Having a single body to accredit assessors and perform other administrative tasks would also be economically efficient.

Given the issues discussed above in relation to rating tools that now form part of legal requirements being operated by commercial, nongovernment entities, it may be preferable if the single agency were a government rather than a private body. However, unless this handover to government were done with the consent of the relevant private body (here the GBCA), it may be difficult to enforce, given the essentially voluntary nature of the GBCA's activities.

COAG 2009 proposes that the new energy standard and rating framework would be implemented through the Building Code of Australia. As the new framework is proposed to address only energy use, the GBCA may continue to administer the broader Green Star rating.

International Movements Toward Standardization

Internationally, the standardization of similar types of tools such as LEED, BREEAM, and Green Star, for example, by using common metrics, is already being discussed (Saunders 2008; UKGBC 2009). This will assist in making comparisons between buildings in different countries and will make it easier to share best practices between tools (Saunders 2008).

The GBCA, its UK and U.S. counterparts, and the agency administering BREEAM recently signed a memorandum of understanding under which they will develop common metrics to measure GHG emissions relating to new homes and buildings, to be used in Green Star, BREEAM, and LEED ratings (UKGBC 2009). The Sustainable Building Alliance, with the backing of several European countries, is also developing common metrics for several indicators, including emissions, though all alliance members will retain their own rating tools. The International Organisation for Standardization is also working on this issue and may prove a useful source of standard metrics.

There do not appear to be specific discussions regarding the standardization of performance tools such as NABERS and Energy Star—although, in fact, such tools may be easier to standardize than design ratings. This may be a fruitful avenue to explore.

Rating Tools Within the Big Picture

In the context of climate change, green building rating tools are commonly seen only as a measure to mitigate climate change, predominantly by reducing energy use. But there is also potential for the tools to assist in promoting adaptation to climate change. For example, design tools could include adaptive features in the categories under which buildings can earn points toward their rating, while performance tools could incorporate a separate category of assessment to measure a building's ability to deal with likely climate change impacts. To some extent, this is already happening (although unintentionally); insulation, which is rewarded under green rating tools for reducing energy use, also reduces vulnerability to weather extremes (Ürge-Vorsatz and Metz 2009). As the effects of climate change become felt more strongly in cities, this option for further development of the tools may receive further attention.

Although green building rating tools can be very effective in encouraging the spread of more energy and resource efficient buildings, without requiring government expenditure, it is also important to recognize their limitations. They are one tool among many and are most effective when accompanied by a series of other measures. One of the most effective of these can be a requirement by certain authorities to use buildings with a certain rating, as is the case under the Energy Efficiency in Government Operations policy.

Financial incentives (for example, matching funding or tax deductions) from government for building owners who retrofit their buildings to achieve certain efficiency standards (which may be set by reference to a green building rating tool) are currently being discussed in Australia and may prove very effective if implemented. However, the cost of such measures to government is often a stumbling block (Larsson 2004). Other types of incentives are possible: Japan has had success with providing preferential planning and development approvals for buildings with high CASBEE ratings, in addition to subsidies and preferential interest rates (Murakami 2009).

In addition, minimum standards set out in a building code (including minimum standards for appliances such as air conditioners) remain important to set the baselines. An effective system for improving energy efficiency in cities would set stringent basic requirements for new equipment, new buildings, and renovations; provide financial incentives for upgrading existing buildings to certain standards; require official building energy ratings to be obtained and disclosed; and also provide encouragement (and reputational benefits) for top-end performance via public recognition of high green building ratings. Japan, which has introduced many of these measures, has shown notable improvement in energy efficiency in the last few decades (Murakami et al. 2009). The Australian measures announced in COAG 2009 also make progress in this direction. Each of these measures should ideally be easy to understand and implement, with information readily accessible online, and each measure should be regularly updated to take account of changing technology, practices, and expectations.

Conclusions

The Australian experience with a design rating tool, Green Star, and a performance rating tool, NABERS, indicates that these tools, although very different, can each play a valuable role in encouraging higher performing buildings, but that the presence of two tools can cause confusion in the marketplace. Some simple measures could be carried out to reduce market confusion and increase standardization of the tools, but government demands for convergence between these tools may raise difficult questions. The useful features of each tool should not be abandoned merely for the sake of convergence, nor should the developers of tools resist all calls for change.

Tools that can be applied relatively quickly and inexpensively to new and existing buildings, that allow for maximum comparability, and that recognize and reward both efficient initial design and efficient ongoing operation are likely to be the most widely used and thus the most effective in reducing emissions. No one tool currently exemplifies all these features, but the movement toward standardization or convergence of tools will mean that, over the next few years, a new generation of rating tools will be developed. Watch this space.

Annex 5A Green Star—Energy Category

Green Star Credit Summary for Energy—Office Design version 3 and Office As Built version 3, last updated March 20, 2009 (available on GBCA Web site, http://www.gbca.org.au/).

Ref. no.	*Title*	*Aim of credit*	*Credit criteria summary*	*No. of points available*
Ene –	Conditional Requirement	To encourage and recognize designs that minimize the greenhouse gas emissions associated with operational energy consumption and maximize potential operational energy efficiency of the base building.	To meet the conditional requirement: The project's predicted greenhouse gas emissions must not exceed 110 $kgCO_2/m^2/$ annum as determined using energy modeling in accordance with: • The Australian Building Greenhouse Rating (ABGR) Validation Protocol for Computer Simulations OR • The final and current version of the Green Star Energy Calculator Guide	
Ene – 1	Greenhouse Gas Emissions	To encourage and recognize designs that minimize greenhouse gas emissions associated with operational energy consumption.	Up to 20 points are awarded where it is demonstrated that the building's predicted greenhouse gas emissions have been further reduced below the Conditional Requirement. No evidence is required in addition to that submitted for Ene– Conditional Requirement.	20
Ene – 2	Energy Sub-metering	To encourage and recognize the installation of energy sub-metering to facilitate ongoing management of energy consumption.	Up to 2 points are awarded as follows: 1 point is awarded where: • It is demonstrated that sub-metering is provided for substantive energy uses within the building (i.e., all energy uses of 100 kVa or greater) and • There is an effective mechanism for monitoring energy consumption data.	1

Ref. no.	Title	Aim of credit	Credit criteria summary	No. of points available
			An additional 1 point is awarded where: • The point above is achieved; • It is demonstrated that sub-metering is provided separately for lighting and separately for power for each floor or tenancy, whichever is smaller; and • There is an effective mechanism for monitoring water consumption data.	1
Ene – 3	Lighting Power Density	To encourage and recognize designs that provide artificial lighting with minimal energy consumption.	Up to 3 points are awarded where it is demonstrated that the lighting power densities for 95% of the NLA meet the following criteria at 720 mm AFFL with the default maintenance factor of 0.8: • 1 point for energy use of 2.5 W/m^2 per 100 Lux; • 2 points for energy use of 2.0 W/m^2 per 100 Lux; and • 3 points for energy use of 1.5 W/m^2 per 100 Lux.	3
Ene – 4	Lighting Zoning	To encourage and recognize lighting design practices that offer greater flexibility for light switching, making it easier to light only occupied areas.	Up to 2 points are awarded as follows: 1 point is awarded where it is demonstrated that: • All individual or enclosed spaces are individually switched; • The size of individually switched lighting zones does not exceed $100m^2$ for 95% of the NLA; and • Switching is clearly labeled and easily accessible by building occupants.	1
			An additional 1 point is awarded where: • The point above is achieved; and • It is demonstrated that an individually addressable lighting system is provided for 90% of the NLA.	1

Ref. no.	*Title*	*Aim of credit*	*Credit criteria summary*	*No. of points available*
Ene – 5	Peak Energy Demand Reduction	To encourage and recognize designs that reduce peak demand on energy supply infrastructure.	Up to 2 points are awarded where it is demonstrated that the building has reduced its peak electrical demand load on electricity infrastructure as follows: 1 point where: • Peak electrical demand is actively reduced by 15% OR • The difference between the peak and average demand does not exceed 40%. 2 points where: • Peak electrical demand is actively reduced by 30% OR • The difference between the peak and average demand does not exceed 20%.	2

Annex 5B Sample Green Star Certificate

Annex 5C Sample NABERS Energy Certificate

NABERS
ENERGY

SAMPLE ONLY

Company ABC

has achieved a **5 star NABERS Energy** tenancy rating for

XYZ building
00 XXXX Street, Sydney 2000 NSW

Rating effective until

March 2010

This Building has:

356 MJ/m² p.a. energy consumption
61 kgCO2/m² p.a. normalised greenhouse emissions
767,911 kgCO2 p.a. actual greenhouse emissions

This Building takes:

18% GreenPower
Rating before GreenPower: 4 stars

This Performance Rating is based on 12 months operational energy use in the Tenancy Building. More stars indicate better energy performance and lower emissions of greenhouse gases. Accredited GreenPower is renewable energy purchased by the building, which has zero greenhouse emissions. NABERS is an initiative of the Commonwealth, New South Wales, Northern Territory, Tasmania, Victoria and Western Australia Governments. It is managed by the Department of Environment and Climate Change NSW Government.

www.nabers.com.au

GreenPower
Accredited Sustainable Energy

NABERS

Annex 5D Draft Building Greenhouse and Energy Performance Certificate

BUILDING ENERGY EFFICIENCY CERTIFICATE

Company name
Building name
101 Something Street
Sydney NSW 2000

Certificate reference number **2008–1234ABC**

Certificate issued by **Department of the Environment, Water, Heritage and the Arts**

Certificate Date **30 October 2008**

Building Information

This certificate covers energy consumption in the base building, with a net lettable area of 12,800 m^2.

NABERS Energy rating

This building has achieved a

4 star NABERS Energy rating

With 5% GreenPower, this rating is increased to 4.5 star
Rating is valid until
30 October 2009

Building performance history

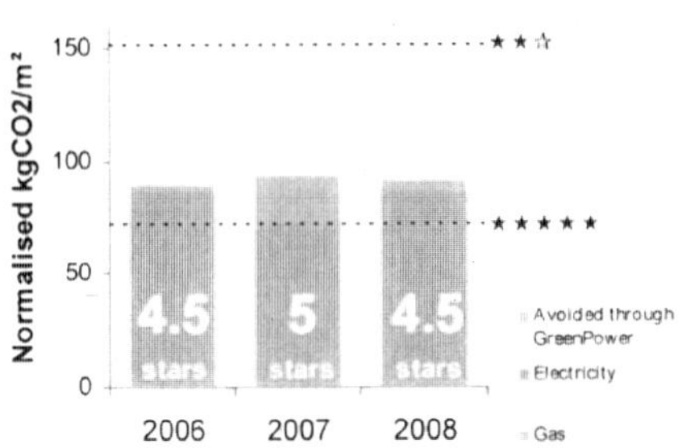

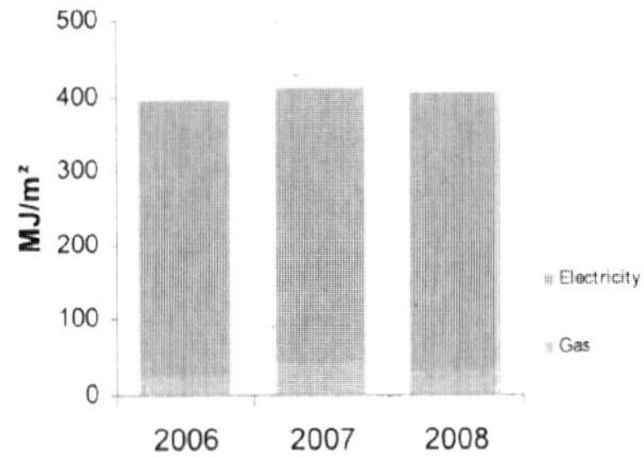

Building energy consumption details

For the period 1 September 2008 – 30 September 2009, this building has:

394 MJ/m² p.a. energy consumption
1,261,298 kgCO_2 p.a. actual greenhouse gas emissions
92 kgCO_2/m² p.a. normalised greenhouse emissions without GreenPower (equivalent to a 4 star rating)
4 kgCO_2/m² p.a. greenhouse emissions offset using GreenPower

Building assessment details

Assessor name:	**John Smith**
Assessor number:	**XYZ012345**
Employer name:	**Energy Engineering**
Trading address:	**12 One St Sydney NSW 2000**
Issue date:	**1 July 2009**

About NABERS

1 Star – Poor energy management or outdated systems
2 Star – Below average building performance
3 Star – Good – above average market performance
4 Star – Very good - strong performance
5 Star – Excellent - best building performance

This performance rating is based on 12 months operational energy use in the Base building only. Please note that if you are intending to sell or lease the rated premises, you may be required to provide a separate tenancy rating certificate to comply with the requirements of the *National Mandatory Disclosure of Building Energy Efficiency Act 200x*.
NABERS is an initiative of the Commonwealth, New South Wales, Northern Territory, Tasmania, Victoria and Western Australia Governments, to rate the environmental performance of Australian buildings. It is managed by the Department of Environment and Climate Change NSW Government.

For more information about how to interpret and use this certificate, visit www.buildingrating.gov.au/commercial.

References

ASBEC (Australian Sustainable Built Environment Council). 2008. "The Second Plank—Building a Low Carbon Economy with Energy Efficient Buildings." ASBEC, Melbourne, Australia.

Australian Greenhouse Office, Department of the Environment and Water Resources, Commonwealth of Australia. 2006. "Energy Efficiency in Government Operations (EEGO) Policy." Australian Greenhouse Office, Canberra.

Baines, T., and J. Bowman. 2008. "Real Estate: Developing Buildings in Cities." In ed. P. Watchman, *Climate Change: A Guide to Carbon Law and Practice*. London: Globe Law and Business.

Campbell, E., and I. Hood. 2006. "Assessment of Tools for Rating the Performance of Existing Buildings: A Report on the Options." Consultant report for Greater Vancouver Regional District." GVRD, Vancouver, BC, Canada. http://www.greenbuildingsbc.com/Portals/GBBC/docs/Assessment_of_Tools_for_Rating_the_Perfomrance_of_Existing_Buildings.pdf, last accessed October 23, 2009.

CIE (Centre for International Economics). 2007. "Capitalizing on the Building Sector's Potential to Lessen the Costs of a Broad-based GHG Emissions Cut." CIE, Canberra, Australia. http://www.thecie.com.au/content/publications/Building_sector's_potential_Sept13.pdf, last accessed October 23, 2009.

City of Sydney Council. 2007. "Draft Ecological Sustainable Development—Development Control Plan." City of Sydney Council, Sydney, Australia. http://www.cityofsydney.nsw.gov.au/Council/documents/OnExhibition/DraftESDDCP/DraftEcologicalSustainableDCP2007.pdf, last accessed October 23, 2009.

COAG (Council of Australian Governments). 2009. "National Strategy on Energy Efficiency." COAG, Canberra, Australia. http://coag.gov.au/coag_meeting_outcomes/2009-07-02/docs/Energy_efficiency_measures_table.pdf, last accessed October 23, 2009.

Cole, R., N. Howard, T. Ikaga, and S. Nibel. 2005. "Building Environmental Assessment Tools: Current and Future Roles." Paper presented at the Sustainable Building 05 conference, Tokyo, Japan, September 28. http://www.sb05.com/homeE.html under "Academic Program," last accessed October 23, 2009.

DECCW (Department of Environment, Climate Change and Water, Government of NSW). 2009. "NABERS Statistics, June 2009." DECCW, Sydney, Australia.

DEWHA (Department of the Environment, Water, Heritage and the Arts, Australian Government). 2008. "Mandatory Disclosure of Commercial Office Building Energy Efficiency—Consultation Regulation Document." DEWHA, Canberra, Australia. http://www.environment.gov.au/settlements/energyefficiency/buildings/publications/pubs/consultation-regulation.pdf, last accessed October 23, 2009.

Evans, P., and T. Wotton. 2008. "Green Buildings and Green Leases—What Is It All About?" *Australian Environment Review* 23 (6): 10-12.

GBCA (Green Building Council of Australia). 2008. "Valuing Green: How Green Buildings Affect Property Values, and Getting the Valuation Method Right." GBCA, Sydney, Australia. http://www.gbca.org.au/docs/NSC0009_Valuing Green.pdf, last accessed October 23, 2009.

———. 2009a. "Green Star Accredited Professional Course Training Notes." GBCA, Sydney, Australia.

———. 2009b. "Green Star Eligibility Criteria." GBCA, Sydney, Australia. http://www.gbca.org.au/uploads/192/960/Green%20Star%20Eligibility%20Cr iteria%20280709.pdf, last accessed October 23, 2009.

Hes, D. 2007. "Effectiveness of 'Green' Building Rating Tools: A Review of Performance." *International Journal of Environmental, Cultural, Economic and Social Sustainability* 3 (4): 143–52.

JLL (Jones Lang LaSalle). 2006. "Assessing the Value of Sustainability." Jones Lang LaSalle, Sydney, Australia. http://www.joneslanglasalle.com/Pages/ResearchDetails .aspx?TopicName=&ItemID=545&ResearchTi tle=Assessing%20the%20Value%20of%20Sustainability, last accessed October 23, 2009.

Larsson, N. 2004. "An Overview of Green Building Rating and Labelling Systems." Paper presented to the Symposium on Green Building Labelling, Hong Kong, China, March 19. http://www.hkpgbc.org/PGBC%20Booklet%20-%20Session %201.pdf, last accessed October 23, 2009.

Lenaghan, N. 2008. "One Green Rating on the Way." *Australian Financial Review,* September 22.

McAteer, A. 2008. "A–Z of NABERS Ratings." Paper presented at the NABERS Benchmark conference, Perth, Australia, May 27.

McKinsey and Company. 2008. "An Australian Cost Curve for Greenhouse Gas Reduction." McKinsey and Company, Sydney, Australia. http://www.mckinsey .com/clientservice/ccsi/pdf/Australian Cost Curve for GHG Reduction.pdf, last accessed October 23, 2009.

Murakami, S. 2009. "Assessment Tools for Building Performance to Promote Energy Efficiency in the Building Sector." Paper presented at the conference on International Standards to Promote Energy Efficiency and Reduce Carbon Emissions, Paris, France, March 16.

Murakami, S., M. D. Levine, H. Yoshino, T. Inoue, T. Ikaga, Y. Shimoda, S. Miura, T. Sera, M. Nishio, Y. Sakamoto, and W. Fujisaki. 2009. "Overview of Energy Consumption and GHG Mitigation Technologies in the Building Sector of Japan." *Energy Efficiency* 2: 179–94.

Nelson, A. 2008. "Globalization and Global Trends in Green Real Estate Investment." RREEF Research Report No.64, RREEF Research, San Francisco, California.

http://www.rreef.com/GLO_en/bin/SO_64_Global_Greening_Trends.pdf, last accessed October 23, 2009.

Ostoja, A. 2008. "ABGR Review—Rating the Rating Tool." Paper presented at the NABERS Benchmark conference, Sydney, Australia, May 8.

Parker, G. 2008. "Overview of the Green Star and NABERS Energy Rating Schemes and Requirements." Paper presented at a seminar of the Law Society of NSW, Sydney, Australia, November 19.

PCA (Property Council of Australia). 2006. "Guide to Office Building Quality."

Saunders, T. 2008. "Discussion Document: Comparing International Environmental Assessment Methods for Buildings." Building Research Establishment, BREEAM International. http://www.breeam.org/page.jsp?id=101, last accessed October 23, 2009.

Smith, T., M. Fischlein, S. Suh, and P. Huelman. 2006. "Green Building Rating Systems: A Comparison of the LEED and Green Globes Systems in the U.S." Carpenters Industrial Council. http://labormanagementcommittee.org/docs/CIC-UMinnGG-LEEDstudy.pdf, last accessed October 23, 2009.

UKGBC (United Kingdom Green Building Council). 2009. Press release: "Common Language for Carbon In Sight: Leading Rating Tool Providers Sign MOU." March 3. http://www.ukgbc.org/site/news/show-news-details?id=132, last accessed October 23, 2009.

Ürge-Vorsatz, D., and B. Metz. 2009. "Energy Efficiency: How Far Does It Get Us in Controlling Climate Change?" *Energy Efficiency* 2: 87–94.

Warren Centre. 2009. "Low Energy High Rise Building Research Study—Final Research Survey Report." Warren Centre, Sydney, Australia. http://www.warren.usyd.edu.au/front_page.html, last accessed October 23, 2009.

CHAPTER 6

Efficient Lighting Market Transformation in the Making—The Philippine Experience

Noel N. Verdote

In 2005, the Philippine government, led by the Department of Energy, implemented the Philippine Efficient Lighting Market Transformation Project (PELMATP) with support from the Global Environmental Facility (GEF) through the United Nations Development Programme (UNDP). Similar to other countries that embarked on technology-specific efficient lighting programs, the Philippines took advantage of this "low-hanging fruit" to enhance and complement its energy security and climate change mitigation programs. Over the years, the program attracted the support of local and national actors in the lighting industry and the market, including those from urban centers. Two years before the official end date of the project, a scaled-up project funded by the Asian Development Bank (ADB) at US$46.5 million—more than tenfold the original GEF grant of US$3.1 million for PELMATP—was implemented; this project was designed to reinforce the initiatives undertaken under PELMATP and to sustain the transition to efficient lighting systems in the country. This chapter assesses the factors influencing a sustained switch to efficient lighting in the Philippines, which are part of the country's "win-win" strategies toward energy independence and climate change mitigation.

Noel N. Verdote is Project Manager of the Department of Energy's Philippine Efficient Lighting Market Transformation Project (PELMATP). The paper that forms this chapter was written for presentation in the Energy Sector Management Program (ESMAP) of the fifth Urban Research Symposium held in Marseille, France, June 28-30, 2009, funded by the World Bank. The ideas, findings, interpretations, and conclusion here, however, do not necessarily reflect those of ESMAP, and all remaining errors are the responsibility of the author The author would like to thank ESMAP of the World Bank for sponsoring the presentation of this paper at the symposium. The author is also indebted to the Philippine Department of Energy (DOE) led by Secretary Angelo T. Reyes, Undersecretary Loreta G. Ayson, PELMATP Policy Advisory Board Chair, and the PELMAT Project Management Office officials and staff, led by Project Director Raquel S. Huliganga, for their valuable assistance during the preparation of this paper.

Introduction

The 21st century is the century of cities, with around half of the world's population already living in cities. The urban population is expected to grow at an even faster rate during the next decades, with the growth largely expected in developing countries. It is projected that by 2050, the urban population in developing countries will total 5.3 billion, with Asia hosting 63 percent (3.3 billion people) and Africa hosting nearly a quarter (1.2 billion people) of the world's urban population (UN–HABITAT 2008). These staggering numbers indicate the increasing contribution of cities and urban centers to the phenomenon of global warming and climate change: as the city populations grow, so do their energy consumption levels that fuel the much needed services for social, economic, and technological development (Jollands 2008). However, by using energy efficiently and thus reducing the level of attendant carbon dioxide (CO_2) emissions, cities potentially can significantly reduce their climate impact. Given the relatively dense concentrations of population and socioeconomic activities in urban conglomerations, Jollands (2008) cites that approximately two-thirds of global energy savings could occur in cities, or possibly even more, considering the role the city can potentially play in implementing national policies aimed at reducing energy consumption. For example, the switch to light emitting diodes (LEDs) for traffic signals and public lighting in Oslo, Norway, and Vaxjo, Sweden, accounted for a 50 to 70 percent reduction in CO_2 emissions from streetlights.

Energy for artificial lighting is an area where savings can be readily, if not easily, realized at the national level, especially in urban centers. Globally, energy for lighting accounts for 19 percent of total electricity use (Rakestraw 2008; IEA 2008) and generates 1.9 billion tonnes of carbon a year, equivalent to three-quarters of the carbon from all car and light vehicle exhaust around the world (McSmith 2006). Experts estimate that energy efficient lighting (EEL) alone, especially the use of compact fluorescent lamps (CFLs), could significantly reduce lighting electricity demand by 40 percent (Rakestraw 2008). During the past two decades, attempts to aggressively promote efficient lighting were introduced as a demand-side management (DSM) initiative. These were mostly interventions by the utilities to induce reductions in peak electricity demand from end users. Some proved to be successful initial steps toward lighting market transformation in countries like Hungary (Ürge-Vorsatz and Hauff 2001), India (IIEC-India 2006), and Thailand (Ratanopas 1997).

However, the shift to efficient lighting by countries, including the Philippines, is not occurring at a satisfactory pace to the nation itself, although the energy saving benefits are obvious. Importers and distributors made insignificant dents in the local market with their efficient lighting products in the 1990s, with price and quality being deterring factors. Skeptics raised eyebrows when those distribution utilities and electric cooperatives complying with the DSM framework of 1996 promoted the switch to efficient lighting. It was not understood why utilities and cooperatives were venturing into an initiative that would reduce their clients' electricity consumption. The private sector, comprising the commercial and industrial sectors, did not shift immediately to using efficient lighting because of the price factor: higher prices were notionally associated with better quality and, often, the more established brands sold higher priced products. Residential consumers likewise were wary about the shift because they did not fully appreciate the benefits associated with EELS. Therefore, the initial Philippine government programs mostly on efficient lighting were focused on information campaigns. One such vehicle was the "Power Patrol" program in the early 1990s, where a switch to the use of fluorescent lamps from incandescent bulbs was among the energy conservation options espoused (Anunciacion 2001). In general, however, consumers were left to experiment on their own as to which efficient lighting products were best suited for their particular needs.

Numerous projects and programs have been implemented on energy efficient lighting in the Philippines. These include a program supported by the U.S. Environmental Protection Agency (USEPA) under the auspices of the Philippines Department of Energy to replace old T12 lamps with the more energy efficient, 32-watt, linear T8 lamps. The International Finance Corporation (IFC) implemented the Efficient Lighting Initiative for CFL program, which utilized seed capital from the Global Environment Facility (GEF). Subsequently, the Palit-Ilaw subprogram under the Philippine government's National Energy Efficiency and Conservation Program was spearheaded by the Department of Energy (DOE), to encourage the shift to CFLs and T8s.

This chapter focuses on the Philippine Efficient Lighting Market Transformation Project, assessing the implementation of strategies under the project to help transform the Philippines into an EEL economy. To be discussed are target-market-specific activities that demonstrate the benefits of switching to the use of EELS, with a focus on applications in urban centers. Also, it is apparent that making a strong case for both energy efficiency and environment preservation is helping make the program acceptable to a wide range of stakeholders. It is practicable that lighting efficiency improvements, with strong support from city and local government executives, will form part of the local solution in the national drive for energy independence/security and will be an effective mitigation tool with respect to climate change.

PELMATP Background

The first privately led, consensus-based program to aggressively promote efficient lighting, initially with the use of CFLs, was the Efficient Lighting Initiative (ELI), mentioned earlier, a three-year, seven-country project that included the Philippines (Birner and Martinot 2003). However, progress made under the ELI project—which provided technical specifications on efficient lighting, DSM templates for utilities/cooperatives that promoted efficient lighting, and other relevant infrastructural support—was largely undone after the project ended. The lack of government ownership of the project, along with public perception that ELI is exclusively for and by the top manufacturers/importers of CFL lighting products, reduced ELI to another short-lived episode of energy efficiency promotions in the country.

Subsequently, in 2004 the Philippines launched the National Energy Efficiency and Conservation Program (NEECP), a five-point framework for achieving energy independence and savings. Under this program, the shift to efficient lighting, starting with the use of slim tubes (T8) as T12 replacements, was advocated primarily to government agencies. To accelerate the pace of transition to efficient lighting, the DOE in 2005 implemented the Philippine Efficient Lighting Market Transformation Project (PELMATP). This program may be regarded as a practical sequel to ELI that builds on the latter's major accomplishments.

PELMATP is a five-year, US$15 million project that receives partial support of US$3.1 million from the GEF through the United Nations Development Programme (UNDP), along with $12 million counterpart contribution from the Philippines or co-financing by some committed government agencies/entities, led by the DOE, and private sector players. According to the DOE and UNDP (2004), "the project addresses the barriers to the widespread utilization of energy efficient lighting systems[1] (EELS) in the Philippines," thereby contributing to the realization of the country's sustainable development objectives and its goals of reducing greenhouse gas (GHG) emissions in the energy sector. A further objective of PELMATP, according to the document, is to accelerate the integration of EEL promotional programs with the energy efficiency and conservation programs (EE&C) of the DOE; enhance the private sector's involvement and appreciation of the benefits of EELS; and ensure that environmental impacts associated with the widespread use of less efficient lighting systems are mitigated. Under the program, the targeted aggregate energy savings over a decade is projected to be 29,000 GWh, an approximately 21 percent reduction from the current Philippines energy efficiency scenario; the equivalent GHG emissions reduction will be about 4,600 Gg of CO_2 (Philippines DOE and UNDP 2004).

Implementation Strategies and Components for PELMATP

Intrinsic to PELMATP is the recognition of barriers that inhibit the adoption of efficient lighting systems across the Philippine residential, commercial, and industrial sectors. Three equally important and reinforcing market transformation elements embody the specific programs that will remove these barriers (figure 6.1). These are the structural, technological, and behavioral changes that together set in motion the transformation process and create the much needed momentum that will help sustain the transformation process beyond the project life, and eventually increase the adoption and widespread use of energy efficient lighting systems in the country. While former U.S. Vice President Al Gore advocated for "new technologies" to realize big change, PELMATP advocates for a modified industry playing field through structural and technological interventions as well as consumer empowerment. These structural and technical interventions, discussed in more detail in this chapter, will pave the way for the institutionalization of the changes

FIGURE 6.1
PELMATP Strategies and Components

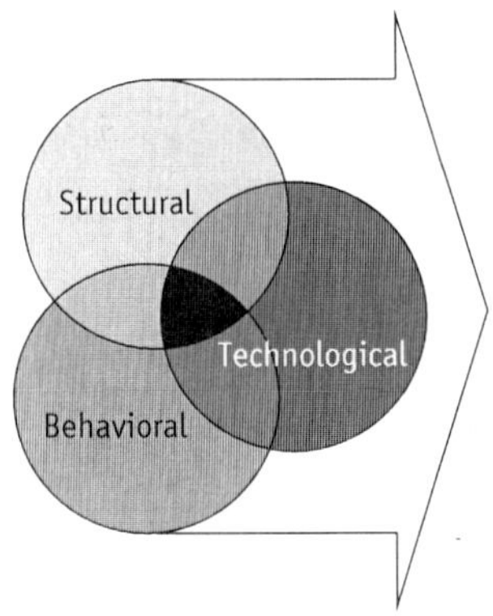

1. EEL policies, standards enhancement program
2. EEL applications institutional and capacity development program
3. EEL applications consumer awareness improvement program
4. EEL initiatives financing assistance program
5. EEL systems waste management assistance program

Source: Philippines DOE and UNDP 2002.

in the transformed industry. On the other hand, empowered consumers making right lighting choices will, at the end of the day, contribute to the desired market outcome.

Barriers and Component Matrix

Several barriers were identified during the PELMATP inception with respect to the adoption of EELS in each of the three sectors—residential, commercial, and industrial—over the five components or main activities of the project (table 6.1). PELMATP instituted programs to remove these barriers focusing on: (1) updating policies, standards, and guidelines on lighting applications; (2) building institutional and technical capabilities; (3) educating consumers and disseminating information; (4) developing and implementing appropriate financing mechanisms; and (5) highlighting the mitigating environmental impacts of the widespread utilization of EELS (figure 6.1).

The PELMATP activities under the five major components were juxtaposed in the framework of market transformation approaches for energy efficient products developed by Birner and Martinot (2003), shown in figure 6.2. These activities have both direct and indirect impacts on the cities' execution of national programs on efficient lighting, as well as those efficient-lighting-related activities best adapted to their local resource conditions. Specifically, PELMATP emphasizes the creation of additional channels of distribution to cater to untapped markets for EELS through utilities, electric and consumer cooperatives, and institutional buyers. These channels have direct contact with a variety of consumers and consumer groups, and their existing infrastructure provisions can readily facilitate the sale of EEL products. Simultaneously, the conventional chain of distribution (shown on the right in figure 6.2) will be strengthened through capacity building by the key players.

TABLE 6.1

Barriers vs. Component Matrix

	Component				
Barriers	*1*	*2*	*3*	*4*	*5*
High initial cost of EEL products	RCI	RCI	RCI	RCI	
Nonimplementation of government incentives	RCI		RCI		
Poor protection of consumers	RCI	RCI			RCI
Poor understanding of use and benefits of EEL		RCI	RCI		RCI
Building designers'/developers' lack of knowledge and simplified tools to calculate full benefits of using EEL products in new commercial establishments	CI	RCI	RCI		
Inadequate promotion and advocacy programs on application of EELS	RCI	RCI	RCI	RCI	RCI
Lack of locally assembled energy efficient luminaires		RCI	RCI		
Poor quality of power supply		RCI	RCI		
Ineffective implementation of the DSM Framework		RCI	RCI	RCI	
Nonimplementation of an outdated Building Energy Use Guidelines	CI		CI		
Inadequate EEL testing facilities	RCI	RCI	RCI		
Insufficient monitoring and verification of products as to their compliance to Philippine National Standards (PNS)	RCI	RCI	RCI		
Poorly developed ESCO transactions		CI	CI	CI	

Source: Philippines DOE and UNDP 2002.
Note: R = Residential; C = Commercial; I = Industrial.

After putting in place all the interventions through the implementation of the barrier removal programs in the course of the five project components, PELMATP developers expect that the majority of untapped market will switch to using EELS. In addition, unlike other EEL market transformation programs, PELMATP also addresses the question of proper management and disposal of lamp waste. Although lamp waste management does not contribute to the overall GHG reduction objective, it was given equal importance with the other components of PELMATP in part because EEL products promoted contain mercury, which means there must be a mandated toxic waste disposal policy.

PELMATP Supply- and Demand-Side Market Transformation Approaches

Birner and Martinot (2003) concluded in their analysis of market transformation approaches for energy efficiency products, particularly for efficient lighting, that interventions should involve both supply and demand sides. As tables 6.2 and 6.3 illustrate, PELMATP engages stakeholders from both sides of the market.

FIGURE 6.2
Market Structure with PELMATP

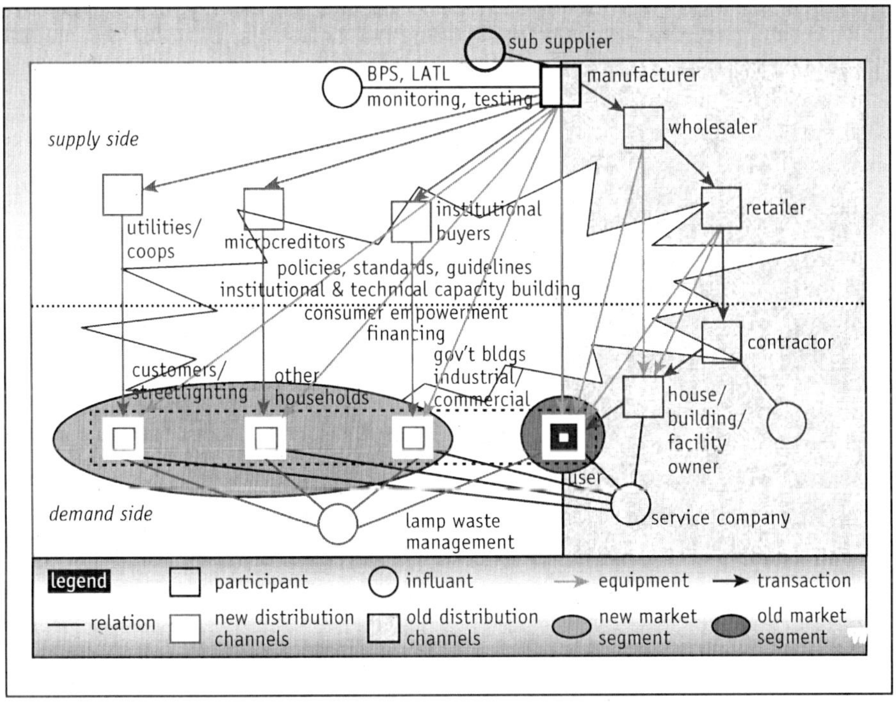

Source: Philippines DOE and UNDP 2002.
Note: To see this figure in color, refer to the appendix at the end of the book.

The tables illustrate the demand-side and supply-side activities that have been initiated under PELMATP to promote the transition toward EELS in the Philippines. However, there appears to be a lack of urgency, along with inadequate political will, with regard to the mandatory implementation of the EEL Standards and Guidelines developed, including those relating to the Philippine National Standards (PNS), Guidelines on Energy Conserving Design of Buildings, Roadway Lighting Guidelines, Eco-labeling Guidelines, Lamp Warranty Guidelines, and Lamp Waste Management Policy. However, it is possible that the awareness of the existence of these enabling mechanisms will create a "market pull" for EELS. Also, the direct involvement of the government in engaging stakeholders, together with the involvement of the private sector and international partners, can potentially accelerate the switch to the use of energy EELS throughout the country.

Significant PELMATP Project Outputs

Significant project outputs from PELMATP laid the foundations of the transition process to efficient lighting systems in cities and among local government units in the Philippines. These project outputs are as follows.

TABLE 6.2
PELMATP Supply-Side Market Transformation Approaches

Item	*Approaches*	*PELMATP activities*
1	Technical assistance and transfer of technical know-how	• Domestic production of lighting products in the Philippines comprises only ballasts and lighting fixtures; PELMATP provides local manufacturing firms technical assistance to enable them to produce EEL products. • PELMATP has involved lighting industry players comprising members of the Philippine Lighting Industry Association (PLIA), as well as non-PLIA members, in an ongoing dialogue and partnership with the government and other entities to promote EEL products.
2	Development of lighting standards and building code	• To date, 25 Philippine National Standards (PNS) on lighting products, including the Minimum Energy Performance Standards (MEPS), have been developed in partnership with the Department of Trade and Industry, Bureau of Product Standards (DTI-BPS). The standards pertaining to CFL are mandatory. • Administrative Order No. 183, crafted under PELMATP and in effect since July 2007, directs the use of EELS in all government offices and establishments. • A "Manual of Practice on Efficient Lighting," originally produced under ELI, and "Guidelines on Energy Conserving Design of Buildings" were updated. "Roadway Lighting Guidelines" have been developed for the first time in the country. PELMATP also helped inform the policy decision on the phase-out of incandescent bulbs by the end of 2009.
3	Voluntary agreements (VA) by private sector	• PELMATP has facilitated the DOE forging voluntary agreements with local suppliers/manufacturers who agree to stock and sell lighting products that exceed the minimum energy performance standards. The DOE executed this in partnership with the Philippine Lighting Industry Association (PLIA) and the DTI-BPS.
4	Incentives for producers and dealers	• Since 2006, PELMATP has recommended the DOE include incentives for manufacturers of EEL products in the government's proposed Investments Priorities Plan (IPP). In 2009, the manufacture of EELS was included in the IPP as among those that will be given incentives (*The Philippine Star* 2009).

5	New distribution mechanisms	• EEL microfinancing models are being implemented through consumer cooperatives throughout the country. • EEL leasing through utilities catering to small commercial and industrial facilities is being worked out with the Cagayan Electric Power and Light Company (CEPALCO). Partnerships between the consumer cooperative and lighting supplier, as well as the utility and the lighting supplier, have been arranged. • Bulk purchases are encouraged by PELMATP for large users of EELS, particularly those in the private sector.
6	Quality testing	• The testing capability of DOE's Lighting Appliance Testing Laboratory (LATL) has been enhanced with installation of new state-of-the-art testing equipment and facilities (including the first goniophotometer facility in the country). The laboratory has been ISO accredited for CFL and linear lamp testing. Also, capacity-building programs have been implemented for officials and staff of the laboratory and other entities of the DOE. • Encouraged by the business prospects of EEL quality testing brought about by the transformation exercise, two other private testing laboratories—the Scientific Environmental and Analytical Laboratory Services (SEALS) and the Institute of Integrated Electrical Engineers of the Philippines Foundation, Inc. (IIEEF)—were established in 2006 to supplement existing government laboratories.
7	Financing for manufacturing upgrades	• PELMATP helped in the development and promotion of energy performance contracting by an energy service company (ESCO). Subsequent to the PELMATP initiative, the DOE issued Department Circular No. DC2008-09-0004 on ESCO accreditation. • Business plan templates were made available to local manufacturers needing financing for manufacturing upgrades. These templates are consistent with those used by the local financing institutions.

Sources: Birner and Martinot 2003; Philippines DOE and UNDP 2004, 2007, 2008, 2009.

TABLE 6.3

PELMATP Demand-Side Market Transformation Approaches

Item	*Approaches*	*PELMATP activities*
1	Consumer education	• The continuous conduct of information, education, and communications (IEC) campaigns across all sectors, including future decision makers (students, employees, housewives, government energy officers, employees, and others). • Knowledge management tools have been developed, e.g., guidelines and guidebooks, manual, info kits, lighting calculators, Web site, EEL course modules, and so on. • Part of consumer education are "Palit-Ilaw" (or switch to the use of EELS) activities aggressively promoted by the project in partnership with the PLIA. Normally, the switch to EELS is done in selected areas or locations for demonstration purposes (e.g., public buildings like schools, markets, hospitals, municipal/city halls, etc.). Since the inception of PELMATP, a total of 17 Palit-Ilaw activities have been conducted, excluding those that are part of the service contracts/technical assistance.
2	Media campaigns to increase awareness among consumers	• Joint media campaigns (radio, TV, and print), as well as on the Internet are also used by the program to increase awareness among consumers about EELS.
3	Professional education of practitioners	• PELMATP provides for partnerships with various professional groups and organizations to promote EEL education among design professionals, property managers, designers, engineers, electricians, and so on. An EEL training course module is also being developed for design professionals and practitioners.
4	Bulk purchases or procurement by public agencies	• Regular updates about PNS-compliant lighting products are provided by PELMATP to the government's Procurement Service for guidance related to its centralized procurement of quality lighting products.
5	Consumer financing (through banks, utility bills, and so on)	• Microfinancing through consumer cooperatives is being formalized. Leasing through utilities is being worked out. • PELMATP trained 10 financing institutions from around the country to improve their understanding and appreciation of the benefits of EELS, particularly in relation to evaluating project proposals.
6	Voluntary agreements by commercial and industrial consumers	• A number of large commercial and industrial consumers signed memoranda of agreement with DOE-PELMATP in support of the use of efficient lighting within their firms.
7	Consumer protection guidelines	• Lamp warranty and eco-labeling guidelines have been developed under PELMATP (Philippines DOE and UNDP 2004). These are pending implementation, although they are supported in principle by the local lighting industry players.

Sources: Birner and Martinot 2003; Philippines DOE and UNDP 2004, 2007, 2008, 2009.

Philippine National Standards for Lighting Products

In 2006, 25 PNS, including Minimum Energy Performance Standards (MEPS), were developed in cooperation with Department of Trade and Industry, Bureau of Product Standards (DTI-BPS). A yellow label, or energy label, is presently mandatory for CFLs while the implementing rules and regulations (IRR) or implementing guidelines (IG) for the rest of the EEL products under PELMATP—linear lamps, ballasts, high-intensity discharge lamps, and luminaries—have yet to be finalized. From June 2007 to April 2008, PNS-compliant CFLs increased from 219 to 303 models, a 27.7 percent increase (Philippines DOE and UNDP 2008).

Guidelines on Energy Conserving Design of Buildings and Roadway Lighting, and Manual of Practice on Efficient Lighting

PELMATP, in partnership with a consortium for technical assistance led by the Institute of Integrated Electrical Engineers (IIEE), the Philippine Lighting Industry Association (PLIA), and the Energy Efficiency Practitioners Association of the Philippines (ENPAP), updated the existing "Guidelines for Energy Conserving Design of Buildings and Utility Systems," which were developed in 1992. Apart from updates on efficient lighting requirements, updates were likewise made on the provisions with respect to other utilities and services within the buildings (such as, building envelope, air conditioning, electrical, and service water heating). The "Guidelines on Energy Conserving Design of Buildings" was officially launched in December 2007, along with the updated "Manual of Practice on Efficient Lighting" (originally published by IIEE with funding from ELI) and the "Roadway Lighting Guidelines" that were developed under PELMATP. The guidelines and the manual were subsequently disseminated in six regions of the country in partnership with IIEE and distributed nationwide.

Also, a tripartite memorandum of agreement (MOA) was signed between the DOE, Department of Public Works and Highways (DPWH), and Department of Interior and Local Government (DILG) in April 2008, for the effective implementation of the guidelines. Subsequently, the DILG issued in August 2008 a "Memorandum to Provincial Governors and City and Municipal Mayors," urging them *"to adhere, or to cause adherence to, energy efficient policies and standards, in both private and government-owned buildings."* Simultaneously, officials from the Office of the Building Officials are being trained to implement compliance with the guidelines. It is expected that sizable energy savings can be realized by local government units (LGUs), particularly city governments, on mandating the use of EELS, initially in new building constructions and during renovations. They can also reduce their streetlighting bills and use the savings for socioeconomic priorities if they were to adhere to the provisions of the "Roadway Lighting Guidelines." An example of the use of the guidelines is given in the section "Value-Added Services by Utilities."

Institutionalizing EELS in Government Offices

An administrative order (AO) crafted by PELMATP was signed by the president of the Philippines in July 2007. This AO institutionalizes the use of EELS in government offices, including national government offices, state universities, and government-owned and -controlled corporations. In 2008, a total of 115 government buildings nationwide had implemented EELS. Among these is Ospital ng Maynila, a government hospital in Manila, which was retrofitted in 2008 with T8s from the existing T12s. This reduced the hospital's electricity consumption by 212,000 kWh and saved the hospital pesos 1.76 million a year (Philippines DOE 2008a). Another example is the replacement of incandescent bulbs in the meat and fish sections of Dagonoy Public Market of Manila, where a typical stall has two 100-W bulbs. These incandescent bulbs were replaced with two 23-W, warm white CFLs in 2005, saving these sections of the market pesos 15,600 a year (Philippines DOE and UNDP 2007). Similar changes were made in other public markets, one in Bacolod City and 10 in Makati City, led by the city government itself, both in 2008 (Philippines DOE and UNDP 2009).

Developing Policies and Standards for B.S. in Electrical Engineering Programs

Since 2005, PELMATP had been working with a committee, composed of representatives from the Commission on Higher Education (CHED), Technical Education and Skills Development Authority (TESDA), and the Board of Electrical Engineering, to review and revise the engineering curriculum for the bachelor of science degree in electrical engineering programs in both government and private universities/colleges. The committee has developed new course modules on illumination engineering for senior electrical engineering students and for vocational/technical courses like those for electricians. The former has been piloted in more than 10 colleges and universities nationwide since 2007 and has been revised accordingly. Reference materials for both courses include the "Guidelines on Energy Conserving Design of Buildings," the "Roadway Lighting Guidelines," and the "Manual of Practice on Efficient Lighting."

Providing Consumer Protection Guidelines

PELMATP has also developed lamp warranty and eco-labeling guidelines for EELS. While neither has been practically implemented yet, the latter, comprising eco-labeling for CFLs, linear fluorescent lamps, and electronic ballasts, has already been approved by the Board of Eco-Labeling Program of the Philippines prior to its consideration (ongoing) by the Government Procurement Service.

Developing Value-Added Services by Utilities

The PELMATP project has been working with utilities to enhance their value-added services to consumers. In fact, LGUs themselves, owing to the high streetlighting bills (unpaid in some cases), became interested in working with the project and the utilities to lower their streetlighting bills. A classic example is the City of Cagayan de Oro, which until February 2008, accumulated approximately pesos 17 million in unpaid bills resulting from overdesigned streetlights. It was estimated that in one stretch of the secondary road in one barangay or village, replacing the existing 50 pieces of 250-W, high-pressure sodium (HPS) with the same number of 70-W HPS could save the LGU around pesos 0.25 million a year (Philippines DOE 2008b). Subsequent to the partnership among the city government, the Cagayan Electric Power and Light Company (CEPALCO), and the DOE-PELMATP, the city government issued an ordinance on streetlighting (City Ordinance No. 10931-2008) following the provisions of the "Roadway Lighting Guidelines" developed under PELMATP. The ordinance is now a ready template for other LGUs to emulate.

Beyond PELMATP

Two years before the end of PELMATP's project life, the DOE has taken a more proactive and aggressive stance in promoting efficient lighting. A "Switch Movement" was launched in July 2008, with participants including members of the government, academia, religious organizations, nongovernmental organizations, social and civic organizations, among others. The movement has issued five calls to action, the topmost being the call to switch to the use of EELS, inspired by PELMATP. The movement aims to mobilize a massive switch to EELS in all sectors of the society, and various LGUs in Manila have already complied with the movement's objective.

Also, the DOE has developed the Philippine Energy Efficiency Project (PEEP), a two-year project that started 2009, with US$46.5 million funding from the Asian Development Bank (ADB). PEEP builds upon the successes of and addresses the gaps identified by PELMATP. The project components of PEEP include the nationwide distribution of 13 million CFLs to households (in exchange for incandescent bulbs), the retrofitting of government buildings/offices, and the transition to LED traffic signals in metropolitan areas. A lamp waste recycling facility will also be installed under this program.

Conclusions and Recommendations

Currently in its fifth and final year, and consistently guided by thematic intervention programs, PELMATP has been instrumental in helping make the Philippines an energy efficient lighting economy. The switch to EELS is seen as one of the easier solutions that will increase in the near term the country's energy security and help mitigate climate impacts through reduced CO_2 emissions. These are important goals because the Philippines is heavily reliant on imported oil and coal for power generation.

EEL solutions can be replicated with ease by LGUs. The implementation of EELS can have a social development dimension, too, because the savings from reduced energy consumption for lighting purposes can be used by LGUs for their socioeconomic objectives. In addition, enforcing the provisions of EEL guidelines can form another source of revenue for LGUs.

For business enterprises the switch to EELS can be a sound investment proposition. Apart from reducing lighting costs in the long term, a greater appreciation by members of the business community of the direct link between energy efficiency and climate change prevention/mitigation can potentially foster a positive image of those firms among consumers. In this context, the use of EELS is a relatively simple approach to incorporating environmental stewardship in the operational principles of businesses.

For the future, conscientious efforts are needed to comprehensively and systematically monitor and account for both the energy savings realized from transitions to EELS and the attendant reduction in carbon emissions. In addition, cities, as well as their commercial and industrial entities, should be able to present a strong business case on the benefits of switching to EELS that can be disseminated widely at both the national and local levels. The knowledge management tools developed by the PELMATP project should be among the information disseminated.

Further, better coordination is necessary between the concerned government agencies associated with the promotion of efficient lighting as part of the local solution to energy security and climate change mitigation. These include the DOE, DTI, DILG, Energy Regulatory Commission, Department of Environment and Natural Resources, Department of Finance, and Bureau of Investment.

At the least, the Philippine government should incorporate a sense of urgency with respect to the implementation of EELS and should synchronize related programs and activities to reinforce each other.

Note

1. The EELS referred to in this project are the energy efficient versions of linear fluorescent lamps (slim tubes), compact fluorescent lamps (CFLs), high-intensity discharge lamps (HIDs), ballasts (low-loss electromagnetic and electronic), and luminaires.

References

Anunciacion, J. 2001. "The Power Patrol Story." http://www.doe.gov.ph/efficiency/power_patrol.htm, last accessed March 23, 2009.

Birner, S., and Martinot, E. 2003. "Market Transformation for Energy-Efficient Products: Lessons from Programs in Developing Countries." http://www.martinot.info/Birner_Martinot_EP.pdf, last accessed April 8, 2009.

IEA (International Energy Agency). 2008. "'Light's Labour's Lost'—Policies for Energy-efficient Lighting." IEA press release, June 29, 2006. http://www.iea.org/textbase/press/pressdetail.asp?PRESS_REL_ID=182, last accessed April 15, 2009.

IIEC-India (International Institute for Energy Conservation—India). 2006. "Energy Conservation and Commercialization II (ECO-II): Support to the Bureau of Energy Efficiency (BEE) Action Plan, Task B4, Program Design Report 'BESCOM Efficient Lighting Program.'" http://www.usaid.gov/in/Pdfs/Bescom_Energy.pdf, last accessed April 3, 2009.

Jollands, N. 2008. "Cities and Energy—A Discussion Paper." In *Proceedings, Competitive Cities and Climate Change, 2nd Annual Meeting of the OECD Roundtable Strategy for Urban Development.* Milan, Italy, October 9–10, 2008. http://www.oecd.org/dataoecd/23/46/41440153.pdf, last accessed January 25, 2010.

McSmith, A. 2006. "A Bright Idea: How Changing Light Bulbs Helps Beat Global Warming (and Cut Bills)," http://www.commondreams.org/headlines06/0703-05.htm, last accessed March 10, 2009.

Philippine Star, The. 2009. "Board of Investment's Notice on General Policies and Specific Guidelines to Implement the 2009 Investment Priorities Plan (IPP)." May 21.

Philippines Department of Energy (DOE). 2008a. "Ospital ng Maynila Energy Audit Report by Project Management Office (PMO)." Consultant report, DOE, Manila.

———. 2008b. *"Pilot Streetlighting Study Results."* Institute of Integrated Electrical Engineers consultant report, DOE, Manila.

Philippines Department of Energy (DOE) and United Nations Development Programme (UNDP). 2002. "Project Brief—Philippine Efficient Lighting Market Transformation Project." Geosphere Technologies, Inc., consultant report, DOE, Manila.

———. 2004. "Project Document—Philippine Efficient Lighting Market Transformation Project." Geosphere Technologies, Inc., consultant report, DOE, Manila.

———. 2007. "Annual Project Report—Project Implementation Review." Project Management Office, consultant report, DOE, Manila.

———. 2008. "Annual Project Report—Project Implementation Review." Project Management Office, consultant report, DOE, Manila.

———. 2009. "Annual Project Report—Project Implementation Review." Project Management Office, consultant report, DOE, Manila.

Rakestraw, D. 2008. "Compact Fluorescent Lamps Could Nearly Halve Global Lighting Demand for Electricity." http://www.lswn.it/en/press_releases/2008/compact_fluorescent_lamps_could_nearly_halve_global_lighting_demand_for_electricity, last accessed April 10, 2009.

Ratanopas. S. 1997. "Success of Demand Side Management Program in Thailand," NEPO World Energy Council. http://www.eppo.go.th/inter/wec/int-WEC-TD02.html, last accessed March 31, 2009.

UN-HABITAT (United Nations Human Settlements Programme). 2008. "State of the World's Cities 2008/2009: Harmonious Cities." London: Earthscan. http://www.unhabitat.org/content.asp?catid=7&cid=5964&subMenuId=0&typeid=46, last accessed January 25, 2010.

Ürge-Vorsatz, D., and Hauff, J. 2001. "Drivers of Market Transformation: Analysis of the Hungarian Lighting Success Story," *Energy Policy* 29 (10): 801–10.

CHAPTER 7

The Role of Information and Communication Technologies for Demand Responsive Transport

Robert Clavel, Elodie Castex, and Didier Josselin

Demand responsive transport (DRT) is a public transport service that provides the user with the advantages of both collective transport and taxi services. Until recently, it was considered a marginal transport mode, reserved for areas with low population density. Since the end of the 1990s, however, the number of DRTs has been increasing consistently, with new investments in urban, suburban, and rural spaces, and with varying degrees of operational flexibility. The flexibility and efficiency of a DRT system are influenced by several factors, the most important being technological, particularly in information and communication technologies (ICT). This chapter illustrates, with two concrete examples, the use of technology to improve DRT efficiency. The type and level of ICT used depends mainly on the type of DRT service, its level of flexibility, and the specific optimization problem needing to be solved.

Introduction

Originating in the mid-1970s, "demand responsive transport" (DRT) services were initially aimed at serving areas with low population density and offering an alternative form of transportation to disabled people. More recently, the use of DRT services has expanded to urban and suburban areas and to offering a variety of services, ranging from regular commuter transport in areas with low passenger volume to transport to specific areas such as airports. The expansion of suburbs and the dispersion of origin and destination points led to the emergence of DRT services.

The general purpose of DRT is to provide public transport service where conventional services would be too expensive and where mobility needs are dispersed,

Robert Clavel is Project Manager on innovative transport for the Centre for the Study of Urban Planning, Transportation and Public Facilities (CERTU) (France). Elodie Castex is Associate Professor at University Lille Nord de France (Lille 1). Didier Josselin is a Researcher at Avignon University (France). This chapter was among the papers selected for the ESMAP-sponsored sessions at the fifth World Bank Urban Research Symposium, but was not able to be publicly presented.

such as during hours of low demand, in areas with low population, and for target users dispersed among the general population (for example, disabled, elderly, students, and tourists).

DRT offers an innovative approach to collective transport in terms of service provision and population targeted. DRT services are more or less flexible depending on the public, the area, and the goals of the service. DRT adapts to the demand of passengers, who must book their trips; the transport service is not provided on a fixed line but is offered in a defined area. Thus, the bus trips are not bound to a specific route or fixed timetable as are conventional services, and they provide flexibility to respond to the demand level and characteristics of passengers. Trips are planned based on user requests specifying start/arrival time and origin/destination. By stopping only at prearranged points at prearranged times, useless trips and passenger waits are avoided. Different categories of DRT have evolved for targeted users: "door-to-door";[1] fixed route; fixed route with deviations; and free route among a set of points, such as "stop-to-stop" services (Burkhardt, Hamby, and McGavock 1995; Ambrosino, Nelson, and Romnazzo 2004; Castex 2007).

In many European cities and regions, DRT services complement conventional and scheduled public transport and offer many advantages and benefits. One of the main reasons for the emergence and success of DRT services is the availability of different information and communication technologies (ICT), which have radically improved the capabilities to provide *personalized* transport services in terms of interface with potential customers, optimization and assignment to meet travel requests, and service provision and management. Adequate ICT support is particularly necessary for the management of booking volumes (number of users), number of trips, and so forth.

This chapter first provides an overview of DRTs in France, then demonstrates how technological opportunities can contribute to the management and development of DRT services, and concludes with examples of the use of ICT in DRT systems.

State of the Art of DRT in France

In France, DRT services are managed by transport authorities that correspond to governmental administrative divisions, such as communes[2] and county councils; sometimes associations or private firms may manage DRTs. In certain cases, a transport authority might manage several services of DRT in different parts of the geographical area.

A database of 650 DRT services in metropolitan France was collected between 2003 and 2005 (Castex 2007) which includes a national survey on the DRT (DATAR, DTT, and ADEME 2004; UTP 2005), as well as information collected from Web sites or directly from transport authorities. The objective of creating the database is to provide an inventory of DRT services at the scale of the "service" rather than at the scale of "transport authority," as in other databases. This provides information on the evolution of specific DRT services and helps map DRT services for the country.

DRT Growth Since the Late 1990s

In 2005, France had 615 DRT services managed by 384 transport authorities and covering more than 7,000 communes. Figure 7.1 presents the number of new DRT services created each year and registered in the database. Their number has increased in France noticeably since the end of the 1990s, encouraged in part by several laws oriented toward better management of transportation systems in accordance with the needs of urbanization and environmental protection. DRT services are especially promoted in cities or territories that lack public transportation. These DRTS offer a variety of services, such as services to targeted users. The majority of DRTs provide transportation services to all the people, like the other modes of public transport (general DRT), while some services are dedicated to specific users such as disabled people (paratransit) or students, customers of private firms (private DRT), members of associations (Social DRT), or railway users (TAXITER).

DRT Services Throughout France

Figure 7.2 shows that DRT services are scattered across France. DRT services for all users (general DRT), many of which are located in rural areas, are more numerous than the others. Social DRTs, only for members of relevant associations, serve relatively few people but are widely distributed in rural areas. TAXITER, taxis service specifically for railway users, is also used in rural areas where rail travel is not common. Many cities use general DRT to complete their transportation systems, for

FIGURE 7.1

Number of New DRT Services Created in France Each Year, 1969–2004

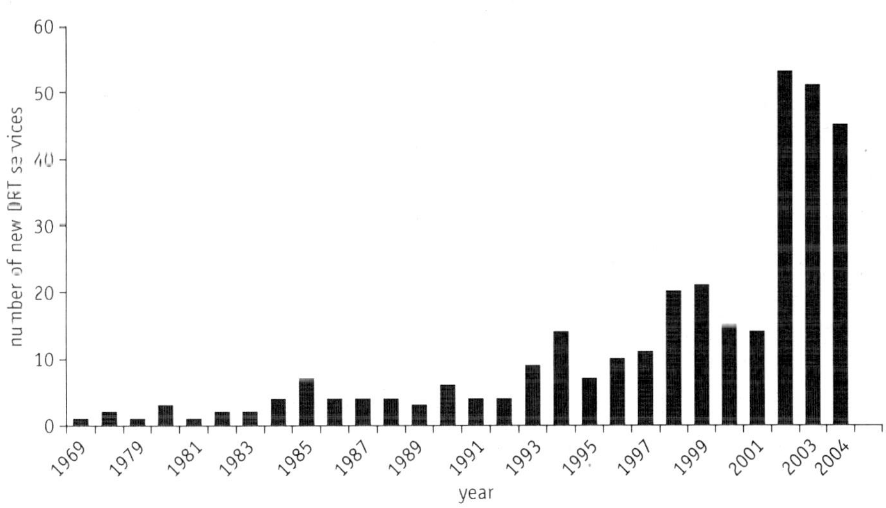

Source: Castex 2007.

FIGURE 7.2
DRT Types and Locations in France, 2007

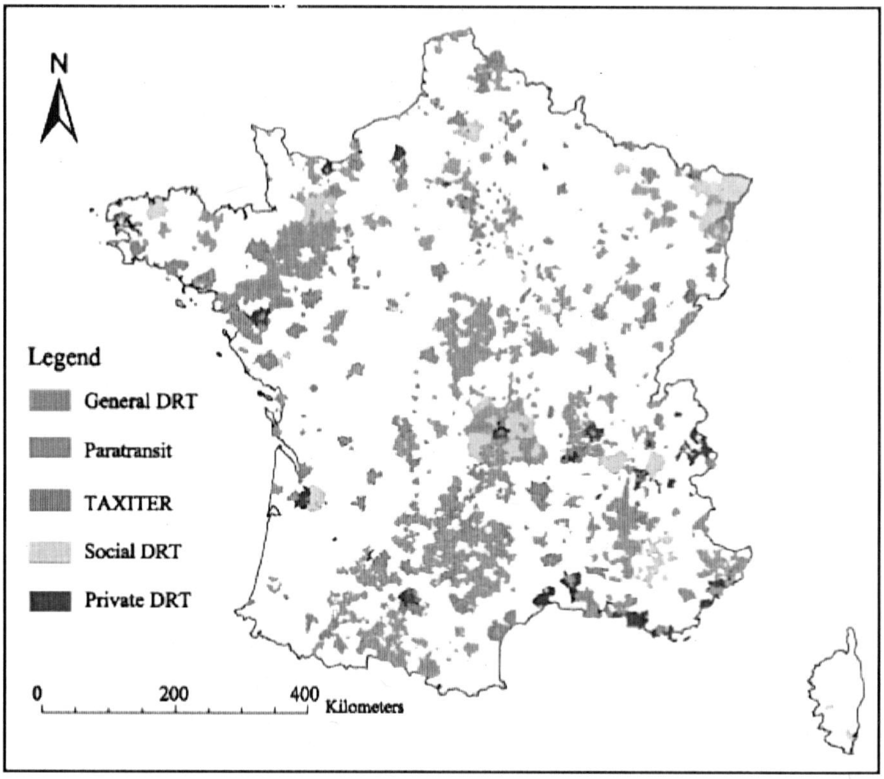

Source: Castex 2007.
Note: To see this figure in color, refer to the appendix at the end of the book.

example, to serve the outskirts of the bus network or to take the place of bus routes during off-peak hours or at night. They are also numerous in suburbs where the transportation network is less efficient. General DRT services represent only 15 percent of France's geographical area but 50 percent of the population. Most cities have also established paratransit DRT services, which provide disabled people with door-to-door services. Private DRT services are usually found only in larger cities. Together, the different types of DRT services cover 24.4 percent of the country, an area in which 90 percent of the French population lives (Castex 2007).

DRT services also differ in supply or availability, offering users varying degrees of flexibility. "Door-to-door" services impose the least constraints, with service comparable to private cars or taxis. At the other end of the spectrum, "fixed route" services are comparable to a bus line: the trip is predefined with given departure and arrival hours. The other types of DRT services offer intermediate levels of flexibility: "stop-to-stop" services permit free routes among a set of points, but their flexibility depends on the number and location of stops; convergent DRT users have their arrival point prefixed but their departure point free. These two DRT service types are relatively similar to door-to-door services in terms of flexibility, while "fixed route with deviations" services are less flexible and relatively similar in characteristics to fixed routes DRT.

Figure 7.3 shows that door-to-door systems are the most widely prevalent, especially with respect to services dedicated to specific users. Many general DRT services are convergent, that is, their destination stops are predefined by the transport authority, but fixed route DRT services are also numerous. Stop-to-stop services and fixed route with deviations are found less frequently in France, although their use has increased over time.

Small-Scale DRT Services in France

DRT services in France are mostly small-scale operations with small numbers of passengers. Indeed, the majority of services are located in smaller administrative areas. Some rural DRT services organized by county councils are larger operations, but still serve areas with low population density. Urban DRT services are usually located in suburbs, where population density is also low. Moreover, the spatial configurations of DRT services rarely correspond to the busiest commuting routes.

Hardware and software innovations have been developed to provide information and communication technology (ICT) support for multi-modal DRT services, including vehicle-locating systems, communications, and networks, and they have generally been developed for specific operators or functions. The main customers for

FIGURE 7.3
Prevalence of DRT Service Types in France, 2007

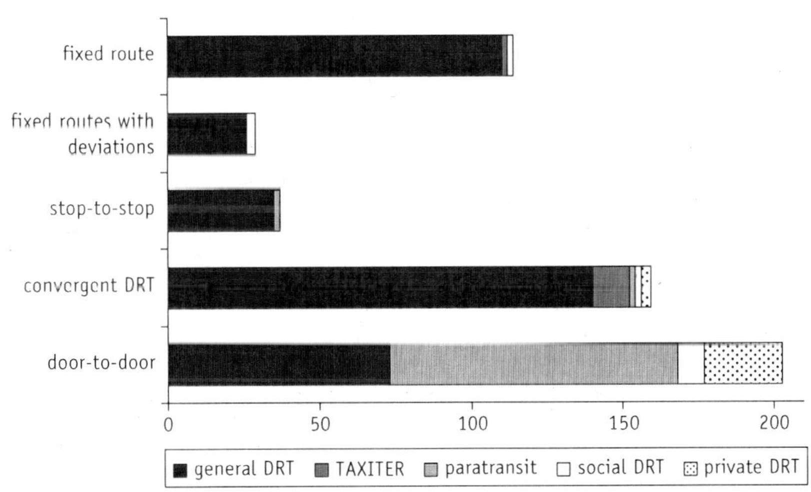

Source: Castex 2007.

such products are public transport operators, regional authorities, and system integrators, who often seek a mix of technologies depending on the operational and financial strategies to be implemented. Although some applications have been developed specifically for managing DRT services, they are few and have not been widely tested. Therefore, there is a strong potential market for ICT products to support DRT technologies.

For the future, DRT services should be enhanced to a larger scale to improve their efficiency, encourage technological innovations integrating their functional requirements, and extend their services to areas with higher population density and more intense commuting requirements (for example, during rush hours). The purpose of large-scale DRT services is to offer a better quality, more competitive service and to realize economies of scale (Tuomisto and Tainio 2005). They also contribute to a green environment by offering a substitute for private cars.

DRTs in Other European Nations

Several large-scale DRT services are available in Europe—Flexlinjen in Göteborg (Sweden), Treintaxi and Reggiotaxi in the Netherlands (Enoch et al. 2004), and Drintaxi in Genoa (Italy), which is a European project under the CIVITAS initiative.[3] Technological improvements have enabled these services to be relatively flexible. Multi-country European DRT services developed in the field of European experimentations, such as SAMPO, SAMPLUS, FAMS[4] (Enoch et al. 2004) and CIVITAS, use computer technology in travel dispatch centers and for communications systems, including on-board systems. Several DRT user interface systems are used via phone, Internet, and GSM/SMS (Global System for Mobile Communications/short messaging service).

SAMPLUS is a project that has demonstrated and evaluated DRT services using ICT technologies at five sites in four European Union (EU) member states (Belgium, Finland, Italy, and Sweden). This project showed that the level of information technology support for DRT is an important factor for operating large-scale DRTs (with high levels of demand, flexibility, and so forth). Unless an operator can confidently predict high patronage and can afford a major investment in hardware and software, it is recommended, as shown by the experience of SAMPLUS, for example, that low to moderate technology solutions are developed. Large-scale investment is most feasible in regulated market environments where more resources, including manpower, are readily available. SAMPLUS also shows that public transport users may regard DRT services as a means of improving intermodality and system integration, especially where there is no such preexisting service, thus opening up mobility opportunities for all citizens and moving one stage closer to seamless public transport. More specifically, DRT services can be tailored to suit the requirements of the local situation, either through highly flexible routing or by guaranteeing connections with conventional services. While it is not the objective of DRT services to adopt a dominant role in the provision of public transport, policy makers should regard it as a vital supplier of services where conventional solutions are untenable (such as in areas with low demand for public transport). Thus, awareness-raising efforts should not only be directed toward local authorities and operators, but also toward central governmental institutions that exert considerable influence over the actions of local authorities.

New Technologies Enhance DRT Flexibility

Flexibility is an important characteristic for any transportation mode. A private car provides great flexibility compared to public transport modes, since it does not involve predetermined trip bookings and allows for door-to-door trips with easy accessibility. However, in European cities, automobile use is increasingly constrained by traffic jams and limited parking availability, especially in city centers. Therefore, public transport systems are more efficient in these areas, although users criticize their lack of flexibility and the numerous connections needing to be made between them because of their fixed itineraries. In more sparsely populated suburbs and smaller towns, the private car still offers the most effective mode of transport. The DRT as an intermediate between private car and bus provides a level of service flexibility that varies depending on the choices of the relevant transport authority.

Defining Flexibility in DRT Systems

Several factors influence the flexibility of a DRT service, including prerequisites for booking a trip and the nature of the trip route. Figure 7.4 compares and evaluates levels of flexibility in DRT services. Each axis represents the main features that influence flexibility in different modes of transport, which can also be applied in the context of DRT functioning. Placement of a DRT service at the center of the graph symbolizes a low level of flexibility, and flexibility increases as the service moves outward from the center. Flexibility is assessed in terms of temporal accessibility (represented in the bottom half of the figure) and spatial accessibility (represented in the top half of the figure, in grey).

For instance, in the top half of the diagram, the axis "direction of flow" indicates flexibility levels in trip opportunities available with a service. A DRT can have "multidirectional" flows (all trips are possible in a given area) or "convergent" flows (only trips toward a convergent point—such as a railway station, town, or shopping center—are possible), with the first being more flexible. "Multi-convergent" flows represent intermediate levels of flexibility, where several convergent points may be available. To the right of "direction of flow" is the axis representing the "spatial functioning of a DRT": a service can have "zonal" functioning (for example, door-to-door), or be organized by "stops" (for example, stop-to-stop services) or on "lines" (for example, fixed route or fixed route with deviations). The "spatial cover" axis (to the left) represents the fact that a DRT service can cover all the territory under the administration of a transport authority or a segment of it. The top three axes

FIGURE 7.4
The Eight Components of DRT Flexibility

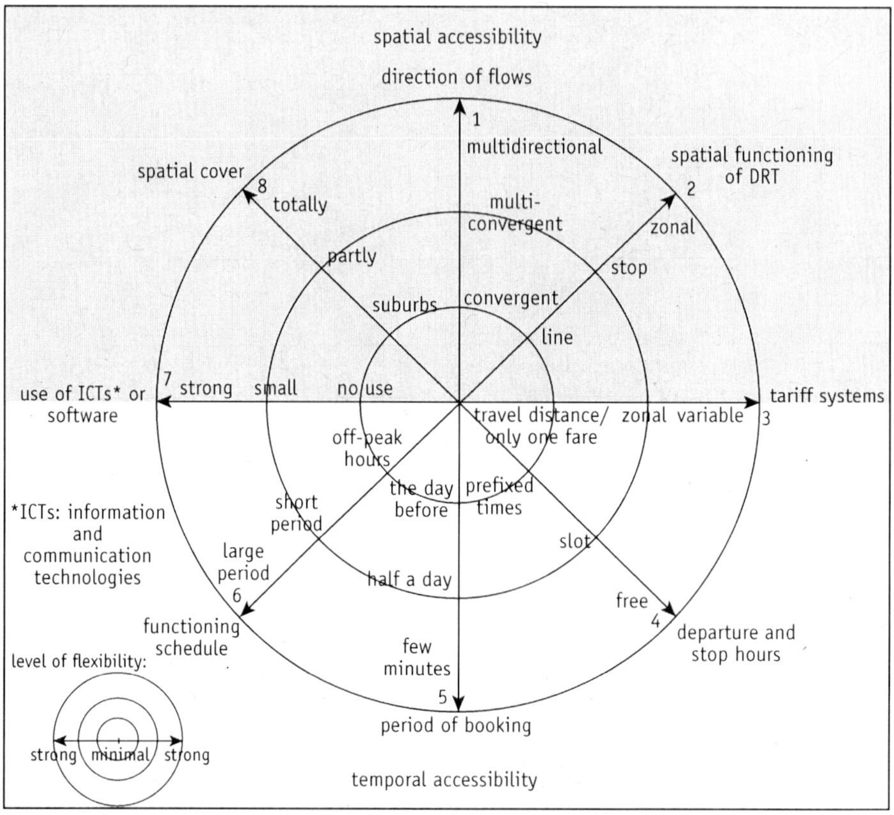

Source: Castex 2007.

together indicate the "spatial accessibility" of a DRT service. On the other hand, "temporal accessibility" is determined by the following:

- The flexibility of the "departure and stop hours": they can be "'free,'' "prefixed,'' or limited according to time "slots.''
- The "period of booking,'' which varies from a few minutes to one day.
- The "functioning schedule'': the DRT can offer services for a large duration of the day (for example, 8:00 a.m. to 7:00 p.m.), a relatively short duration (for example, 10:00 a.m. to 5:00 p.m.), or only during off-peak hours.

Between the above two groups of axes (spatial and temporal accessibility) are located the components that depend on both time and space variables. "Tariff systems'' can be based on time (for example, "variable'' tariffs based on peak and off-peak hours) or space (for example, distance covered or a "zonal'' price). "Use of ICTs'' (axis at left, center) includes time and space dimensions, too, by permitting better trip management.

Without ICTs, it is impossible to manage a flexible service in a large area or one intended for a large number of users. Three levels of flexibility with respect to ICT use are represented in figure 7.4 (lower left circle): no use, minimal use, and strong use.

ICT Technology and DRT Services

The success of DRT services is in large part a result of the availability of various information and communication technologies (ICTs) that have radically improved transport authorities' ability to provide personalized transport services in terms of interface with potential customers, optimization of assignment to meet travel requests, and service provision and management. Continued advances in information technology platforms (advanced computer architecture, Web platforms, palmtops, PDAs, in-vehicle terminals, and so on) and in mobile communication networks and devices (GSM, GPRS, GPS, and so on) have supported the following innovations:

- DRT operators have been supported with respect to service model dimensions, including route, timing of services, and vehicle assignment, and can more readily alter the service offered in response to current or changing demand.
- ICTs have made easier such tasks as trip booking, user trip parameters, negotiation phases, communication of trips to drivers, service follow-up/location, and reporting the completed service.

ICT-based computer architectures supporting DRT operations are usually organized around the concept of a travel dispatch center (TDC). The TDC is the main technological and organizational component supporting the management of DRT services. Computer architecture elements developed for DRT operations include the following:

- Several integrated software procedures to support the operations and management of DRT TDCs, including technological developments to ease procedures for handling user requests, trip booking, service planning, vehicle dispatch, vehicle communications and location notification, systems data management, and regular public transport notification.
- A communication system, usually based on public or private long-range wireless telecommunication networks, supporting communication and information exchanges (both data and voice) between the TDC and the DRT vehicles.
- Several types of DRT user interfaces that enable communication between the passenger and the TDC through different channels. These user devices include phone, Internet, GSM/SMS, and automated answering devices such as IVR (interactive voice response) with CTI (computer telephone integration), which allows the TDC to identify and gather information from the passenger.
- Onboard systems, such as in-vehicle terminals (IVT), installed on DRT vehicles to provide driver support functions during vehicle operation in the form dynamic journey information, route variations, passenger information, and driver/dispatcher messages.

All existing DRT installations are designed with variations in the above basic computer architecture schemes, the implementation of which is made possible by a number of key enabling technologies. These technologies are as follows:

- Booking and reservation systems to manage customer requests.
- Regular public transport information (dynamic or static) meant for DRT operations that support regular public transport services, or to avoid conflicts of DRT schedules with regular public transport schedules.
- Web, IVR systems with CTI, and hand-held devices to assist customer booking.
- Dispatching software for allocating trips and optimizing resources.
- Communication network to link the TDC with drivers and customers.
- In-vehicle units to support the driver.
- GPS-based vehicle location systems.
- Smart-card-based fare collection systems.

The role of the TDC is important for maintenance of system performance and service provision (figure 7.5). The optimized management of user requests through the TDC with the help of ICT also leads to a potentially more economically efficient model of DRT operations.

FIGURE 7.5
DRT System Architecture

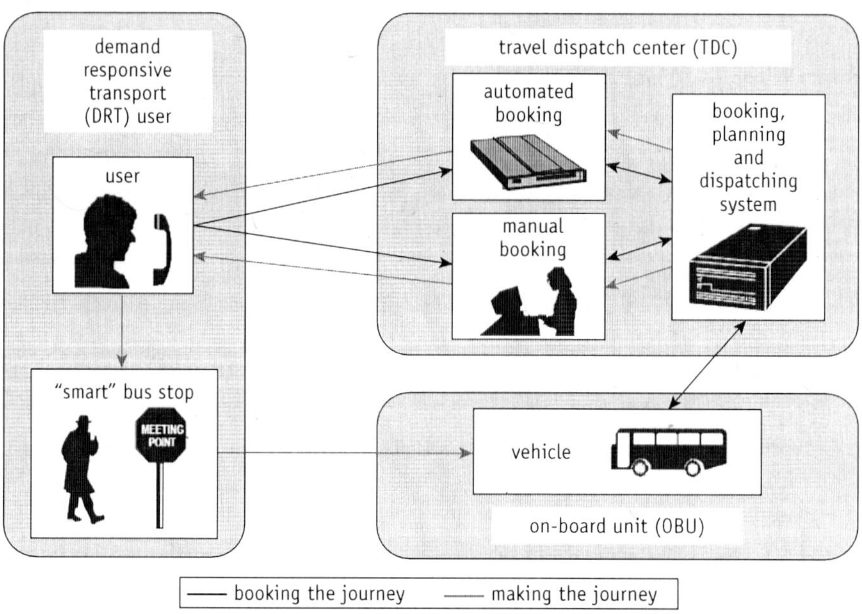

Source: Ambrosino, Nelson, and Romnazzo 2004.

As shown in figure 7.5, the ICT provides the following functions:

- The way the user contacts the TDC.
- The way the TDC integrates all the information about the user's trip request and dispatches it to the vehicle operator.
- The way the operator receives the trip information for all users and organizes the trip.

Examples of DRT Services Using Specific Types of ICT

TADOU: an Innovative DRT in a Rural Area

The "Pays du Doubs Central" is a group of 99 communes across five little towns located in the northeast of France. To serve the mobility requirements of the sparsely populated area of 25,000 inhabitants (Castex 2007), the transport authority developed a stop-to-stop DRT service named TADOU (Transport à la Demande du Doubs Central) using the Tadvance network.[5]

A set of stops cover the entire territory in a system of multidirectional flows, and passengers can travel in any direction from one stop to another, after having phoned the previous day to book their trip. The DRT service uses a software innovation called GaleopSys, developed by Tadvance members, which calculates the trips booked and minimizes the distances traveled by grouping passenger trips as closely as possible. The input data requirements of Galeopsys are stop points, timetables, and the acceptable levels of delay-time.

GaleopSys achieves trip optimization using a geographic information system (GIS), the database of which contains information on all the stops and passenger addresses (figure 7.6). The software enables the calculation of the shortest routes while maximizing the number of passengers transported, in order to use as few vehicles as possible. At the same time, it ensures that users' time constraints are respected. The algorithm developed is based on Dial A Ride Problem with Time-Windows (DARPTW) (Garaix et al. 2007); Prorentsoft distributes the software.

The cost of travel depends on the distance traveled and is proportionally less per mile for a longer distance than a shorter one. In 2006, the number of trips was 1,863 for 2,454 passengers, which corresponds to 1.3 passengers per vehicle. At the beginning of 2007, the service provided 230 trips a month (Castex 2007).

Toulouse Private DRT Service Complements Public Transport Facilities

A DRT service called TAD 106 (Transport à la Demande 106) was created in Toulouse, France, in 2004, by the Public Transport Authority, Tisseo. RCSmobility (Réseaux, Conseils et Solutions Informatiques, or RCSI), the operator,[6] is a private company

FIGURE 7.6

Trip Optimization Using GaleopSys Software (TADOU)

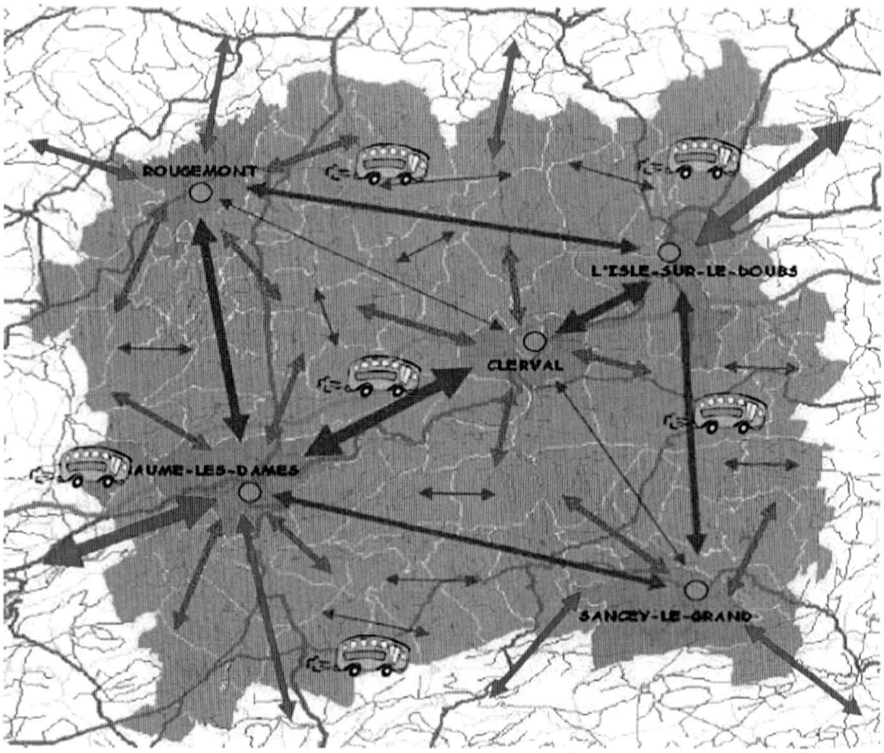

Source: Castex 2007.
Note: To see this figure in color, refer to the appendix at the end of the book.

that provides the eastern suburban area of Toulouse with a flexible public transport service that is complementary to existing regular lines. RCSI developed the ICT innovations[7] used for this operation, which center on a TDC operated by telephone and secured reservation Web site with a multimodal database. The database is regularly updated by SYNTHESE[8] software for the reservation and dispatching of trips. SYNTHESE is free software with a general public license. Its input data requirements are the stop points by zone, "rendez-vous" timetables, "trip duration contract" by pair of departure and arrival stops, the window of time allowed for prior booking (one hour in the TAD 106 line, no reservation is needed if the trip starts at the subway station), and preexisting rules of "no competition" with regular public transport lines. The dispatcher can consult the real-time location of DRT vehicles thanks to GPS and radio coupling, and can inform passengers in case of vehicle delay. As roadmap trips are transmitted automatically to the drivers from the reservation center, operating costs are economized. There is also a permanent radio link between vehicles and the information and reservation call center, and the system displays real-time information about DRT departures; this time is available on screens at the Balma Gramont station, the terminal metro station. Computer terminals at the station can print tickets for the allocated trips. Different-sized DRT vehicles can carry 8 to 22 passengers. Illustrating the success of this initiative are the statistics associated with this DRT service: on average, 950 passengers per day were transported in 2009, with more than 1,400 passengers per day during special events like the music festival; a total of 295,000 trips were made in 2009, a 95 percent increase from 2006, with a passenger satisfaction rate of 97 percent.[9]

Many outstanding features are associated with TAD 106 (figure 7.7), including the following:

- The coverage of a large geographical scale: trips are possible on every origin/destination among 100 stop points across six cities.
- A direct connection to a fast major transport mode (subway).
- A high level of availability, with departures every 30 minutes from 5:00 a.m. to 12:30 a.m. every day of the year.
- Flexible operations that require no preestablished itinerary—only stops and timetables are fixed.

FIGURE 7.7

Principles of DRT TAD 106 in Toulouse

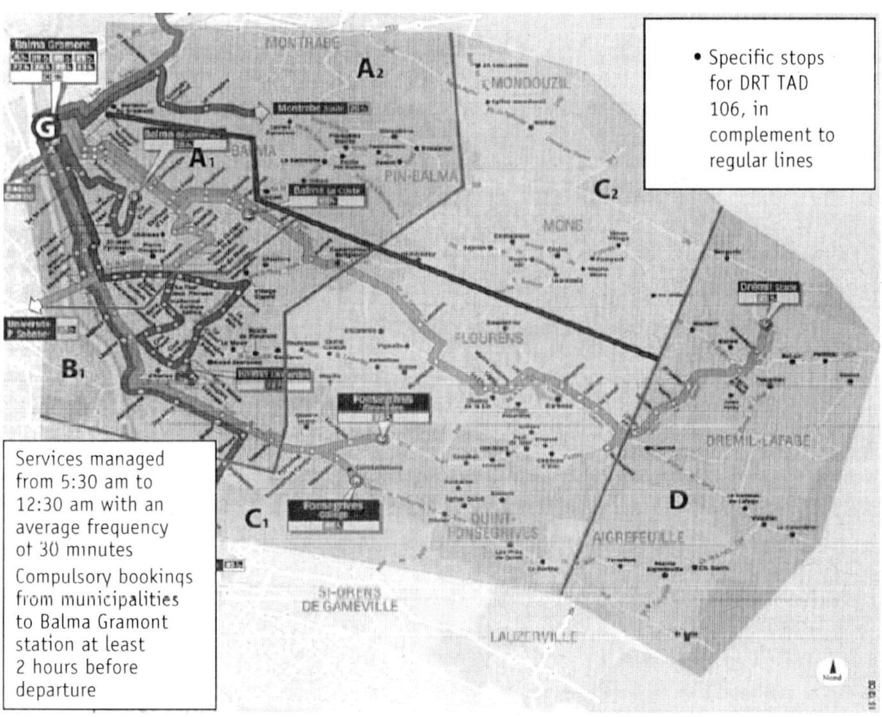

Source: DRT operator in Toulouse.
Note: To see this figure in color, refer to the appendix at the end of the book.

- An adaptable system allowing for variations in the number of DRT vehicles plying the roads as a function of demand.
- A low-constraint travel option for passengers, since it provides the option for booking an unplanned return trip from the central area of town at the metro terminal; otherwise, bookings are to be made one hour in advance, with cancellation possible until departure.
- Accessible information and booking center, with special toll-free telephone lines available between 6:30 a.m. and 10:30 p.m. every day of the year. In addition, a secured Web site allows customers to book places every time.
- The DRT service operates to complement the existing urban public transport network; there is no competition with regular lines and a common system of information is shared with them, with integrated tariff systems, and so forth.

It is envisaged that future ICT innovations in several realms (see figure 7.8) will allow improvements in the quality of DRT services being offered in Toulouse and will lead to the reduction of management costs. These new technologies will depend to a large extent on the will of the entities involved, including the local authority and the operators, and will benefit from the integrated workings of communication satellite systems with local telecommunications networks.

FIGURE 7.8

Technological Perspectives for DRT TAD 106 in Toulouse

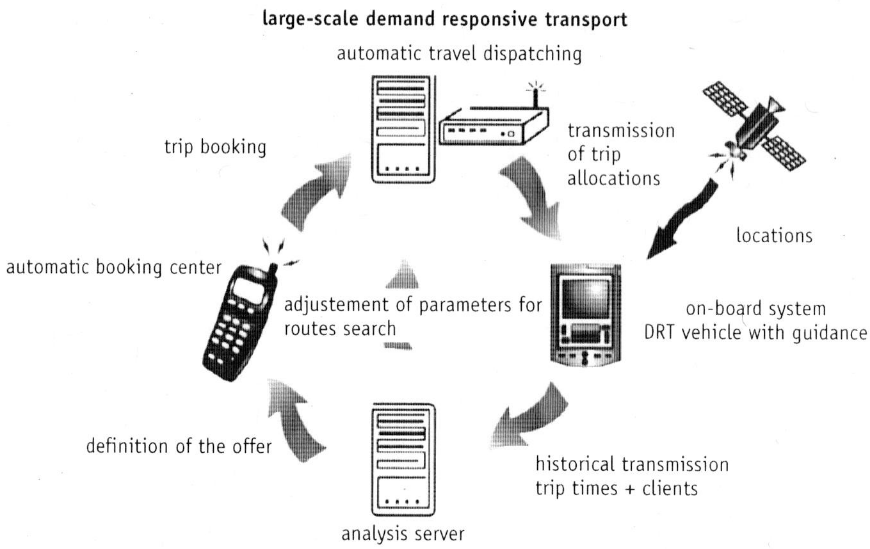

Source: Toulouse DRT operator (RCSI).

Environmental Impacts of DRT

An ADEME (French Environment and Energy Management Agency) analysis (ADEME 2005) compares greenhouse gas (GHG) emissions arising from the use of urban DRT services against a theoretical situation where a passenger would use a private car for the same trip. The study concluded that:

- DRT services consume slightly more energy (10 percent more at maximum), and therefore emit more GHGs when compared with the usage of private cars.
- DRT operations consume much less energy than regular public transport services (at least 60 percent less) in areas with low mobility potential, for example, those with low population densities.

Although the comparison shows a slightly higher level of energy consumption, and corresponding GHG emissions, using a private car for the same trip, the results must also take into account that many trips would simply not be made without the availability of DRT services. This is the case because one of the objectives of the DRT model is to provide mobility to those who find it difficult to use both private and public forms of transport, such as the elderly, disabled, and those who do not own personal transport and/or have inadequate access to public transport.

In addition to the findings above, a simulation-based assessment was also conducted on the extension of the DRT service "Evolis Gare," which provides connecting trips to the train station for people living in Besançon municipality. The simulation consisted of providing trips from the entire urban agglomeration of Besançon, including suburban areas (Castex 2007). This assessment, however, showed that a DRT service with a high rate of vehicle occupancy (at least three people/vehicle) enables a decrease in the distance traveled (up to 30 percent fewer kilometers) by DRT vehicles. Therefore, if trips are optimized, DRT services can contribute to a reduction in GHG emissions from vehicles.

Conclusions

One of the major reasons for the spread of DRT in recent years has been technology developments. Advances in software, computers, digital maps, expert systems, remote communications, in-vehicle computers, and GPS technologies have helped make DRT services logistically and economically viable. The examples of Doubs Central and Toulouse, two different areas of France, show that technology can play a key role in optimizing DRT trips and bringing quality service to the population in a large area, especially when the patronage is high.

Technology offers the potential for almost "real-time" demand responsiveness in transport services, particularly in complex networks, to a level far more advanced than manual systems. However, the costs of establishing high-tech schemes are significant, which can result in reluctance among local authorities to make the required software investments. Moreover, suppliers often require specialized hardware, rather than adapting standard platforms, which increases cost considerations and thus constrains the greater use of technology for more efficient DRT services.

There is immense potential for DRT services to develop as an economically sustainable public transport means and an alternative to the private car, in particular to meet the travel needs for target groups of passengers such as the elderly, disabled, and other special groups. These potential markets have largely not been met by transport services because cost-effective means have not been adequately developed. Operators and local authorities increasingly believe that if technical barriers can be overcome, the transport market for DRT will accelerate.

The environmental benefits of an efficiently functioning DRT system have not been adequately predicted as yet. However, it is conceivable that if DRT systems can use ICT developments to achieve more flexibility in their response to mobility requests, and are utilized to capacity, they can lead to a reduction in transport-related GHG emissions by limiting private vehicle usage to a large extent.

Notes

1. "A service that picks up passengers at the door of their place of origin and delivers them to the door of their destination" (Burkhardt, Hamby, and Gavock 1995).

2. The commune is the smallest administrative subdivision in France, numbering 36,000 in metropolitan France.

3. CIVITAS (**CI**ty-**VITA**lity-**S**ustainability) is a European Community initiative aimed at supporting and evaluating the implementation of ambitious integrated sustainable urban transport strategies in European cities (see http://www.civitas-initiative.org).

4. Experimental DRT systems in Italy, Finland, Belgium, Sweden, and England for citizens in rural and urban areas.

5. A network of researchers of the universities of Avignon and Belfort-Montbéliard aimed at the development of DRT.

6. Public transportation innovative company, located in France and Switzerland, providing services to network authorities (free software development, consulting, call center, data typing, and so forth.

7. Developed by RCSmobility.

8. The TAD 106 uses the reservation module of SYNTHESE, the most advanced open-source multimodal traveler information system.

9. Data collected from Tisséo, Public Transport Authority.

References

ADEME (French Environment and Energy Management Agency). 2005. "Fiches Descriptives de Transport à la Demande pour Flux Faibles." Study conducted by ISIS for ADEME.

Ambrosino, G., J. D. Nelson, and M. Romnazzo. 2004. *Demand Responsive Transport Services: Towards the Flexible Mobility Agency.* Rome, Italy: ENEA.

Burkhardt, J., B. Hamby, and A. T. McGavock. 1995. "Users' Manual for Assessing Service-Delivery Systems for Rural Passenger Transportation." TCRP Report 6, National Academy Press, Washington, DC.

Castex, E. 2007. *Le Transport a la Demande (TAD) en France: de l'État des Lieux à l'Anticipation. Modélisation des Caractéristiques Fonctionnelles des TAD pour Développer les Modes Flexibles de Demain.* Ph.D. thesis in Geography, University of Avignon, France.

DATAR (Delegation for Country Planning), DTT (National Director for Ground Transport), and ADEME (French Environment and Energy Management Agency). 2004. *Services à la Demande Innovants en Milieu Rural: de l'Inventaire à la Valorisation des Expériences.* Study by ADETEC for DATAR, DTT, ADEME.

Enoch, M., S. Potter, G. Parkhurst, and M. Smith 2004. *INTERMODE: Innovations in Demand Responsive Transport.* Final Report, Department for Transport and Greater Manchester Passenger Transport Executive. London: Department for Transport. Available from http://www.dft.gov.uk.

Garaix, T., D. Josselin, D. Feillet, C. Artigues, and E. Castex. 2007. "Transport à la Demande en Points à Points en Zone Peu Dense, Proposition d'une Méthode d'Optimisation des Tournées." *Cybergeo* 396: 17.

Tuomisto, J. T., and M. Tainio. 2005. "An Economic Way of Reducing Health, Environmental, and Other Pressures of Urban Traffic: A Decision Analysis on Trip Aggregation." *BMC Public Health* 2005: 14.

UTP (Union des Transports Publics Ferroviaires). 2005. *Le Transport à la Demande dans le Transport Public Urbain.* UTP publication available at http://www.utp.fr/publications.

CHAPTER 8

Getting to Carbon Neutral: A Review of Best Practices in Infrastructure Strategy

Christopher Kennedy, David Bristow, Sybil Derrible, Eugene Mohareb, Sheyda Saneinejad, Robert Stupka, Lorraine Sugar, Ryan Zizzo, and Bernard McIntyre

Measures of cost-effectiveness for reducing greenhouse gas (GHG) emissions from cities are established for 22 case studies, mainly involving changes to infrastructure. GHG emissions from cities are primarily related to transportation, energy use in buildings, electricity supply, and waste. A variety of strategies for reducing emissions are examined through case studies ranging from US$15,000 to US$460 million of capital investment. The case studies have been collected to support a Guide for Canadian Municipalities on Getting to Carbon Neutral (G2CN). The cost-effectiveness, given by annual GHG emissions saved per dollar of capital investment, is found to vary between 3 and 2,780 t e CO_2 (tonnes of carbon dioxide equivalents) per year per US$ million for the G2CN database. The average cost-effectiveness of the database of 550 t e CO_2 per year per US$ million is significantly exceeded by solid waste projects in Canada (Federation of Canadian Municipalities); and by developing world projects under the Clean Development Mechanism (CDM). Five case studies in the G2CN database with GHG savings over 100,000 t e CO_2 are highlighted. Yet, cities need to start planning projects with reductions of the order of more than 1 million t e CO_2 per year in order to substantially reduce emissions below current levels for smaller cities (1 million people or fewer) and mega-cities.

Introduction

The main sources of greenhouse gas (GHG) emissions attributable to cities are transportation, energy use in buildings, electricity supply, and, to a lesser extent, waste

Christopher Kennedy, David Bristow, Sybil Derrible, Eugene Mohareb, Sheyda Saneinejad, Robert Stupka, Lorraine Sugar, and Ryan Zizzo are with the Department of Civil Engineering, University of Toronto, Ontario, Canada. Bernard McIntyre is with the Toronto and Region Conservation Authority, Ontario, Canada. Financial support for this work was provided by Toronto and Region Conservation Authority, Toronto, Canada.

(Kennedy et al. 2009). Transportation emissions per capita are inversely related to urban density; sprawling, low-density cities designed around automobiles have higher emissions than more compact cities with substantial public transportation. Building energy use is primarily dependent on climate, that is, heating degree days, but can also be impacted by the quality of building envelopes. Emissions from electricity depend to some extent on the level of consumption, but more significant is the means of power generation; nuclear and renewable sources (hydro, solar, wind, and so on) have close to zero direct emissions. Emissions from landfill waste, which are often particularly significant for cities in the developing world, are primarily dependent on the extent to which waste methane or other gases are captured. Overall, it is clear that urban GHG emissions are highly dependent on a range of infrastructure systems.

In order to reduce GHG emissions, it is necessary to first understand their scale. Inventories of emissions, primarily from developed cities, typically vary from about 3 t e CO_2 (tonnes of carbon dioxide equivalents) per capita to over 20 t e CO_2 per capita (Kennedy et al. 2009). This means that for mega-cities of close to 10 million people, GHG emissions are on the order of 100 million t e CO_2. For example, Los Angeles County has estimated emissions of 122 million t e CO_2 and New York City has emissions of 85 million t e CO_2 (table 8.1). Smaller western cities, of 1 million people, typically have GHG emissions on the order of 10 million t e CO_2.

Many different infrastructure strategies can be employed in reducing GHG emissions from cities. Table 8.2 displays a range of strategies under categories of transportation/land-use, buildings, energy supply, solid waste, water/wastewater, and carbon sequestration. Most of these strategies involve changes to infrastructure or built form, but a few economic strategies, such as congestion pricing, are also considered. The strategies are also classified in terms of their scale of engagement (that is, the size of the project). Those strategies with higher scales of engagement generally entail higher investments and produce greater GHG reductions (relative to strategies

TABLE 8.1

Total GHG Emissions from Ten Global Urban Regions

Urban region	*Year*	*Population*	*GHG Emissions (million t e CO2 /yr)*
Los Angeles County	2000	9,519,338	122.0
New York City	2005	8,170,000	85.3
Greater London Area	2003	7,364,100	70.6
Greater Toronto Area	2005	5,555,912	63.4
Bangkok (City)	2005	5,658,953	60.1
Cape Town (City)	2005	3,497,097	40.1
Denver (City and County)	2005	579,744	12.3
Greater Prague Region	2005	1,181,610	11.0
Barcelona (City)	2006	1,605,602	6.7
Geneva (Canton)	2005	432,058	3.3

Source: Adapted from Kennedy 2009.

TABLE 8.2

Preliminary Classification of GHG Reduction Strategies by Scale of Engagement

Category	*Minor*	*Medium*	*Major*
Transportation/land-use	• High-occupancy vehicle lanes, smart commute, carpool networks, car share • Natural gas vehicles (e.g., municipal buses) • Bus rapid transit • On-road bike lanes • Bike share	• Financial penalties to auto use (e.g., tolls, congestion charging) • Incentives for use of low-emission vehicles • Light rail transit • Segregated bike lanes	• Pedestrianization of city centers • Infrastructure for plug-in hybrid electric vehicles • Subways • Bicycle highways
Buildings	• Building energy retrofits • Green roofs • Energy star buildings	• Improved building operations • Photovoltaics • Solar water/air heaters • Ground source heat pumps	• Demolition and reconstruction with high energy-efficiency green buildings
Energy	• Vertical axis wind turbines	• District energy systems • Borehole or aquifer thermal storage	• Nuclear power plants • Concentrating solar generation
Solid Waste	• Landfill methane capture • Vacuum collection of solid waste	• Solid waste gasification	• Increased recycling • Greening supply chains
Water/Wastewater	• Reduced demand through low-flush toilets or low-flow shower heads	• Reduced demand through grey-water systems	• Anaerobic waste water treatment plants
Carbon Sequestration	• Planting of urban forestry • Algae	• Residential scale urban agriculture in CO_2-enriched greenhouses	• Industrial scale urban agriculture in CO_2-enriched greenhouses • Carbon offsets

Source: Kennedy and Mohareb, forthcoming.
Note: The seventh category, integrated community design, cuts across the six categories shown.

in the same row). The designation of scale of engagement in table 8.2 is essentially a preliminary hypothesis for this research, though it is based on 10 years of research and teaching on sustainable urban infrastructure. Clearly, the GHG reductions achieved through building a subway or a concentrating solar plant are much higher than those from a bicycle lane or a vertical axis wind-turbine. The authors' aim is to quantify this scale of engagement more rigorously.

Among the strategies listed in table 8.2 are some that reduce GHG emissions through increased energy efficiency and others that use carbon-free technologies. Strategies such as high-occupancy-vehicle lanes, bus rapid transit, green roofs, and improved building operations make more efficient use of energy, but, in the absence of other strategies, still entail GHG emissions; these are examples of low-carbon alternatives. Other strategies, such as pedestrianization, photovoltaics, wind turbines, and concentrating solar generation, eliminate direct GHG emissions. Further, strategies in table 8.2 can also be emissions-free, depending on conditions. For example, a carbon-free electricity supply will enable the use of light rail transit, subways, electric vehicles, and ground source heat pumps that have no GHG emissions. Employing strategies that use fossil fuel energy more efficiently help cities *approach* carbon neutrality, but to actually reach carbon neutrality substantially requires carbon-free technologies.

Objectives

The specific objectives of this chapter are as follows:

1. Identify which infrastructure strategies can lead to the greatest reductions in GHG emissions from cities in both developed and developing nations.
2. Identify which infrastructure strategies are most cost-effective in reducing GHG emissions (that is, quantify reductions in t e CO_2 per dollar investment).

Researchers addressed these objectives by collecting data from approximately 70 case studies of best practices in carbon-neutral urban design. This case study review is the first phase of a larger project, described in the next section, which aims to produce a guidebook for cities on how to become carbon neutral. The results of the case studies show projects ranging from US$15,000 to US $460 million of investment, with GHG savings between 45 t e CO_2 and 950,000 t e CO_2. Findings from the dataset are compared to those from two other sources: projects under the Federation of Canadian Municipalities' Green Municipal Fund and those funded under the United Nations Clean Development Mechanism (CDM). Caveats to the case study approach are included in the discussion.

Background: Getting to Carbon Neutral

As part of the Toronto Region's Living City initiative (TRCA 2008), the Sustainable Urban Infrastructure Group at the University of Toronto is producing the *Guide for Canadian Municipalities on Getting to Carbon Neutral (G2CN),* a guidebook to assist medium to large Canadian municipalities on the path to becoming carbon neutral (Kennedy 2009). In the analysis undertaken for this guidebook under the G2CN (Getting To Carbon Neutral) project, the authors define carbon neutrality as the requirement that direct and indirect emissions from a municipality, minus sequestered carbon and offsets, should total zero, a definition used in this chapter as well. Getting to a carbon-neutral state will first entail developing low-carbon cities.

As a first step to addressing climate change, many cities have inventoried GHG emissions, often using the simple, pragmatic approach of ICLEI (International Council for Local Environmental Initiatives).[1] In Canada, 157 municipalities are participating in the Partners for Climate Protection (PCP) program. Most of these municipalities have established inventories of GHG emissions; however, many are struggling to develop and implement strategies for substantially reducing GHGs.

Yet there are many examples of sustainable design practices both in Canada and elsewhere that have led to lower GHG emissions for various neighborhoods or infrastructure systems within cities.[2] Many of the strategies employed are substantial, long-term endeavors, requiring serious investment and some degree of societal change. These projects demonstrate that if municipalities were to aggressively pursue a wide range of GHG-emissions-reducing strategies, subject to their own unique conditions, it may be technically feasible for many of them to become carbon neutral.

The first phase of the G2CN project entails collecting and analyzing best case practices and strategies in sustainable urban design and planning. This review includes case studies in transportation, buildings, energy systems, waste management, water infrastructure, carbon sequestration, and integrated community design. The case studies are discussed below.

The second stage of G2CN involves developing best practice strategies for reducing municipal GHG emissions in the categories of buildings, transportation/land-use, energy supply, municipal services, and carbon sequestration/offsets (table 8.3). For each strategy, the guide provides simple, generic rules of thumb for estimating the reductions in GHG emissions that can be achieved. The formulas can be used, for example, to estimate GHG reductions from installing X km of light-rail, constructing a gasification plant to process Y tonnes of solid waste, or servicing Z percent of homes in a municipality using a district energy scheme.[3] The rules of thumb typically calculate changes to intermediary quantities, such as energy use and vehicle kilometers traveled, from which GHG emissions are subsequently determined. The guide does not seek to be prescriptive on how the GHG reduction strategies are selected; instead it offers a menu of choices.

Methodology

In this chapter, data from the case studies assessed in the G2CN guide is used to estimate the cost-effectiveness of GHG emissions-savings initiatives. Cost-effectiveness has been defined as follows:

$$\text{Cost-effectiveness} = \frac{\text{Annual GHG emissions saved}}{\text{Capital investment}}$$

TABLE 8.3

Strategies for Reducing Municipal Greenhouse Gas Emissions

BUILDINGS
Strategy 1: Reduce Energy Demand
Strategy 2: Utilize Solar Energy
Strategy 3: Ground Source Heat Pumps
TRANSPORTATION/LAND USE
Strategy 1: Appropriate Land Use
Strategy 2: Public Transportation
Strategy 3: Active Transportation
Strategy 4: Deter Automobile Use
Strategy 5: Changing Vehicle Technology
ENERGY SUPPLY
Strategy 1: Electricity from Renewable Sources
Strategy 2: Aquifer and Borehole Energy Storage
Strategy 3: District Heating and Cooling
Strategy 4: Combined Heat and Power
WASTE, WATER AND MUNICIPAL SERVICES
Strategy 1: Increased Recycling
Strategy 2: Waste Incineration and Gasification
Strategy 3: Methane Capture
Strategy 4: Water Demand Management
CARBON SEQUESTRATION AND OFFSETS
Strategy 1: Urban Agriculture and CO_2-enriched Greenhouses
Strategy 2: Urban Forestry
Strategy 3: Geological and Mechanical Sequestration

Source: Kennedy 2009.

The above measure only considers initial capital costs; it excludes recurring costs, user fees, financial benefits, low-cost financing, and government subsidies. Given this scope, cost-effectiveness may be considered as a limited economic measure because it gives no indication of financial returns. However, it is useful from the perspective of capital budgeting to reduce GHG emissions.

The case study selection procedure sought to establish leading-edge examples of initiatives being taken by municipalities, cities, or regions to reduce GHG emissions, both in Canada and worldwide. Extensive teaching and research experience on a wide range of sustainable urban design topics provided an initial background to the selection of the case studies.The topics include green buildings (Zachariah, Kennedy, and Pressnail 2002; Dong, Kennedy, and Pressnail 2005; Saiz et al. 2006); urban water systems (Sahely and Kennedy 2007; Racoviceanu et al. 2007); sustainable urban transportation (Kennedy 2002; Kennedy et al. 2005); alternative energy systems (Kikuchi, Bristow, and Kennedy 2009); sustainable neighborhoods (Engel-Yan et al. 2005; Codoban and Kennedy 2008); urban metabolism (Sahely, Dudding, and Kennedy 2003; Kennedy, Cuddihy, and Engal-Yan 2007); and the application of the principles of industrial ecology to the design of sustainable cities (Kennedy 2007).

For each case study, the aim was to establish a database on costs, benefits, barriers to implementation, and GHG savings. Thus, the case studies also provided empirical data to support/verify the *rules of thumb* developed in the Canadian municipalities guide. This chapter uses the data from these case studies to examine the cost-effectiveness of strategies for reducing emissions. The criteria for selection of the case studies were as follows:

- The use of strategies that reduce, or prevent growth of, GHG emissions.
- The coverage of both Canadian and non-Canadian best practices.
- Inclusion of examples from both medium and large municipalities.
- Strategies that primarily focused on technological and urban design solutions, rather than economic measures.
- The availability of information.

The chosen case studies (table 8.4) include a wide range of strategies, essentially covering the categories established in table 8.2.

TABLE 8.4

Capital Costs and Annual Greenhouse Gas Savings for the Case Studies

Project	*Location*	*Capital cost (US$ million)*	*Annual GHG Savings (kt CO_2 e)*
TRANSPORTATION/LAND USE			
Light Rail Transit	Calgary, Alberta, CA	447	591(v)
Rubber-tired streetcar	Caen, FR	279	
Quality Bus Corridor	Dublin, Ireland	68	
Bus Rapid Transit	Vancouver, BC, CA	39.2	1.8
Metrolink: Express Bus	Halifax, NS, CA	9.3(v)	1.125(*)
Heavy-Duty HCNG Transit Buses-Hydrogen Highway	Port Coquitlam, BC, CA	2.3(v)	0.12(v)
Low Emission Zone	London	90(v)	
Congestion Charging	London	244(v)	120(v)
Bike Share	Paris	132(v)	18(*)
Bike Share	Barcelona		1.92
Bike Campaign	Whitehorse, YT, CA	1.5(v)	0.0045(v)
Real time information	Portland, OR, US	6	
High Occupancy Vehicle lanes	Seattle, WA, US	2.3 (v)	
Parking Cash Out	California		0.24(v)
BUILDINGS			
Demolition/Reconstruction	Toronto		31.4 (v)
Solar Air Heating	Montreal	1.96	1.342
Solar Hot Water Heating	Paris	0.91 (v)	0.214 (v)
Ground Source Heat Pump	Concord, Ontario, CA		2.862
Building Integrated Photovoltaic	Coney Island, New York City		0.086

(continued)

TABLE 8.4

Capital Costs and Annual Greenhouse Gas Savings for the Case Studies *(continued)*

Project	*Location*	*Capital cost (US$ million)*	*Annual GHG Savings (kt CO_2e)*
Green Roof	San Francisco	2.64	
Heat Recovery from Restaurant Exhaust	Toronto	0.015	0.0075
ENERGY			
Solar Central Receiver Station	Seville, SP	41	110(*)
Solar Thermal Electricity Plant	Mojave Desert, NV, US		270(*)
Tidal Stream System	Northern Ireland	5.4 (v)	2(v)
Solar Power and Borehole Thermal Storage	Okotoks, Alberta, CA	3.8 (v)	0.26 (v)
Photovoltaic Plant	Olmedilla de Alarcon, Spain	460	29(*)
Wave Power Plant	Portugal	10.6	1.8(*)
Geothermal Power	Northern California		950(*)
Lake Water District Air Conditioning	Toronto		79
Small Hydro	Cordova Mines, Ontario, CA	1.36	0.06(*)
Urban Wind Power	Toronto	1.21	0.38
Vertical Axis Wind	Liverpool, UK	0.46 (v)	0.0014(*)
SOLID WASTE			
Source-Separation and Methane Production	Sydney, Australia	75 (v)	210 (v)
Incineration-Based CHP	Gothenburg, Sweden	453 (v)	205 (v)
Methane Capture	Toronto	24 (v)	
WATER/WASTEWATER			
Biogas from sewage	Stockholm, Sweden	15	14
Co-Generation at Wastewater Treatment Plant	Ottawa, CA	3.4	
Wastewater heat recovery	Sony City, Japan		3.5 (v)
CARBON SEQUESTRATION AND OFFSETS			
Doubling Urban Canopy	Chicago	10/year (v)	170 (v)
SUSTAINABLE COMMUNITY			
Vauban	Freiburg, Germany		2.1
Dockside Green	Victoria, BC, CA	4.5 (v)	5.2 (v)
Dongtan	Shanghai, China		750 (v) expected

Source: Kennedy 2009.

Note: v = verified; * = GHG; calculation undertaken by the project team.

The geographical extent of the case studies is biased first toward Canada and second toward North America and Western Europe. The locations of the case studies are shown in figure 8.1. Clearly, it would be useful to include more examples from other parts of the world, especially Asia.[4]

Information on each case study was first assembled from Web sites describing the infrastructure or other relevant literature. This information was then sent by e-mail to owners, designers, or managers of the infrastructure, who were invited to verify, update, and add to the case study descriptions. Case studies for which the information has been verified are marked with a "v" in table 8.4.

In some cases, information was obtained only on energy saved, or vehicle kilometers reduced, thus requiring the study team to estimate the corresponding GHG savings. For example, if the case study involved electricity supply from a renewable source, the authors established the GHG savings relative to the conventional supply, based on provincial, state, or national GHG intensity as documented, for example, by Ontario Power Authority (OPA 2007) or by the U.S. Energy Information Administration (EIA 2006). Another illustrative example is the calculation of GHG savings related to the MetroLink express bus project in Halifax, Canada. For this calculation, the authors

FIGURE 8.1
Locations of Case Studies for the Getting to Carbon Neutral Project

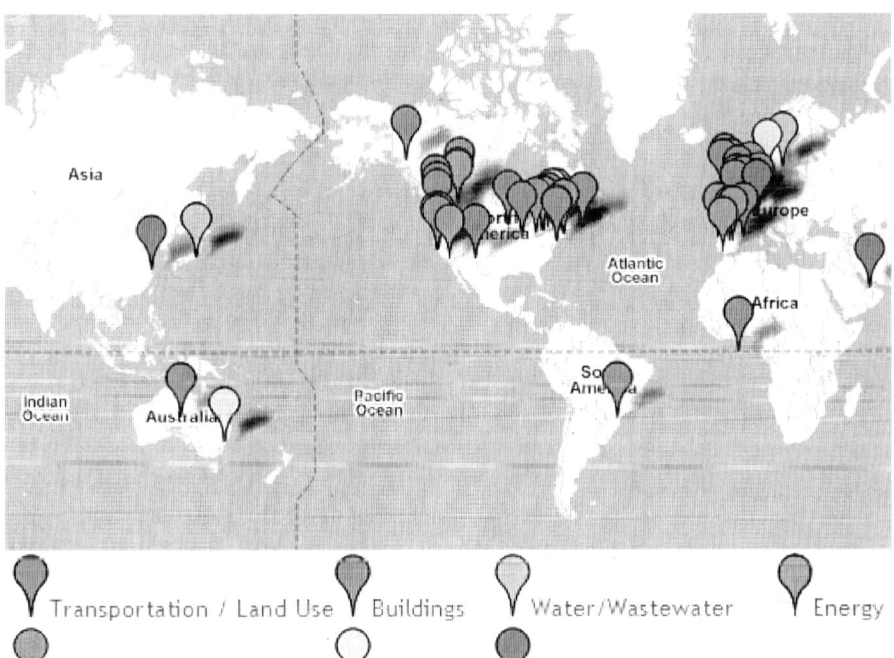

Source: Google map adapted from the project Web site: http://www.utoronto.ca/sig.
Note: To see this figure in color, refer to the appendix at the end of the book.

multiplied the number of round trips made in 2008 by the average daily GHG savings per passenger, assuming that all riders used a private automobile prior to using the MetroLink. Similar assumptions are made by a number of other agencies, for example, the City of Calgary, in calculating and reporting the GHG savings of other case studies presented in this report. Case studies for which the authors estimated GHG savings are marked by an asterisk (*) in table 8.4.

Analysis of Case Studies

From the 68 case studies for which information was sought, data on annual GHG savings and capital costs was obtained for 42 cases. Of these, 34 cases have data on GHG savings, 30 have data on capital costs, and 22 have both sets of data (table 8.4).

For the case studies where the capital costs and GHG emissions are both known, there is a relatively consistent fit of increased emissions savings with higher investments (figure 8.2). However, the data are plotted on a log-log basis, since both the costs and GHG emissions vary over orders of magnitude. The log-log plot disguises the very large deviations in the dataset. For example, a bicycle campaign in Whitehorse, Canada, costing US$1.51 million is estimated to save 45 t e CO_2 per year, while a solar air heating system in Montreal costing US$1.96 million has reported GHG savings of 1,342 t e CO_2 per year. Another comparison can be made between a subway line in Rennes, France, saving 18,000 t e CO_2 per year at a capital cost of US$550 million, and Calgary's (Canada) light rail transit, powered by wind-generated electricity, which saves

FIGURE 8.2

Log–Log Plot of Annual GHG Savings vs. Capital Costs for Infrastructure Case Studies in G2CN Project

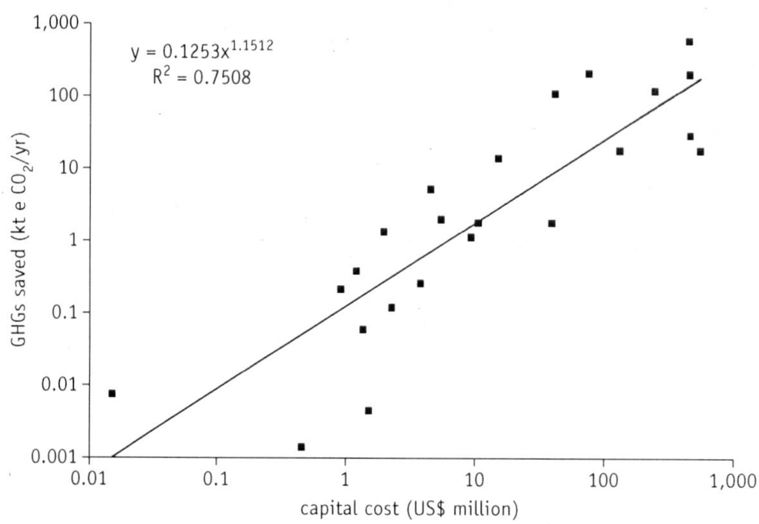

Source: Kennedy 2009.

591,000 t e CO_2 per year, after a capital cost of US$447 million. Clearly, there are significant differences in cost-effectiveness among the case studies with respect to reducing GHG emissions.

While the line of best fit in figure 8.2 is of limited use as a predictor, it helps to distinguish the infrastructure investments achieving the most cost-effective reductions in GHG emissions. Points that lie above the line, in the middle range of costs, include cases of solar hot-water heating, urban wind power, tidal stream power, and generating biogas from sewage, as well as the Montreal solar air heating system.

Five case studies at the top end of figure 8.2 are particularly noteworthy because they lie above the line of best fit and exceed GHG savings of 100,000 t e CO_2 per year. They are the following:[5]

- Seville's Solar Central Receiver Station (box 8.1)
- London's Congestion Charging Scheme (box 8.2)
- Gothenburg's Combined Heat and Power (CHP) System (box 8.3)
- Sydney's Source Separation and Energy Recovery Facility (box 8.4)
- Calgary's Light Rail Transit System (box 8.5)

In addition to these five cases, the dataset includes four other projects with annual GHG savings greater than 100,000 t e CO_2, but for which the capital costs were not available to the study team. They are the following: (1) a solar thermal electricity plant in the Mojave desert (270,000 t e CO_2 per year); (2) a series of more than 20 geothermal power plants in Northern California (950,000 t e CO_2 per year); (3) Chicago's plan to double its tree canopy (170,000 t e CO_2 per year); and (4) the planned Dongtan sustainable community development near Shanghai, China (750,000 t e CO_2 per year). However, it must be mentioned that doubts have been raised on whether the Dongtan development will proceed (*The Economist* 2009).

Box 8.1

PS10 Solar Central Receiver Station, Seville, Spain

With a peak power capacity of 11 MW, Seville's Solar Central Receiver Station is the first commercial grid connected version of its type. The infrastructure consists of 624 120m^2 heliostats that reflect sunlight onto a receiver atop a 100-m tall tower, which produces steam to drive a turbine. The facility produces 24.3 GWh of electricity per year, of which only 12 to 15 percent is provided by backup natural gas. The project cost US$55 million (IEA 2008); and it is estimated that it saves 110,000 t e CO_2 per year. There are plans to expand the system to 300 MW by 2013, which would be enough to power 180,000 homes, that is, approximately the entire city of Seville.

Source: Abengoa Solar 2008.

Box 8.2

Congestion Charging, London, United Kingdom

Vehicles that drive within a clearly defined zone of central London between the hours of 7:00 a.m. and 6:00 p.m., Monday to Friday, have to pay an £8 daily "congestion charge." Payment of the charge allows drivers to enter, drive within, and exit the "charging zone" as many times as necessary on that day. The congestion charge was first introduced in Central London in February 2003, with the daily charge of £5 per day to travel between 7:00 a.m. and 6:30 p.m. In July 2005, the charge rose to £8, and the zone was extended in February 2007, when the hours of operation were reduced. There is no charge for driving on the boundary roads around the zone. In addition there are a number of routes that enable vehicles to cross the zone during charging hours without paying—the Westway (raised highway) and a route running north to south through the center of the zone.

If the congestion charge is not paid, a penalty charge notice (PCN) for £120 is issued to the registered keeper of the vehicle. This is reduced to £60 if paid within 14 days; but if a PCN is not paid within 28 days, the penalty increases to £180. Net revenues raised from congestion charging are spent on improving transport in London. In 2007–08, the scheme generated a net revenue of £137 million (Transport for London 2008).

London's Congestion Charging Scheme is estimated to have reduced emissions by 120,000 t e CO_2 per year; it cost about US$324 million to implement, including traffic management measures, communications/public information, systems set-up, and management.

Source: Evans 2008.

Box 8.3

CHP from Solid Waste, Gothenburg, Sweden

Gothenburg's Combined Heat and Power (CHP) system, fueled by waste incineration, reduces municipal solid waste disposal needs and displaces fossil-fuel-generated heat and electricity. Approximately 1.2 million MWh of electricity were produced from incineration of waste in 2006. Annual benefits from the sale of electricity are US$33.6 million. Other benefits may include avoided landfill disposal costs and carbon credits. The system cost US$600 million and saves about 205,000 t e CO_2 per year through separation and combustion of degradable organic carbon.

Source: Climate Leadership Group 2008a.

The case studies mentioned above with savings of more than 100,000 t e CO_2 per year cover a variety of sectors: transportation, solid waste, energy generation, and urban forestry. This trend is encouraging, as it shows that a diverse range of effective strategies can be adopted to reduce GHG emissions. For some of the nine case studies,

Box 8.4

Source Separation and Energy Recovery, Sydney, Australia

Sydney's Source Separation and Energy Recovery Facility achieves a 70 percent diversion of municipal solid waste from landfill. An anaerobic digestion process produces methane, which is combusted to produce electricity to power the separation facility. Compost from the organic stream is sold for US$20–US$30 per tonne. The estimated GHG savings are 210,000 t e CO2 per year, as landfill gas emissions are avoided through separation/combustion of degradable organic carbon. The facility was constructed at a capital cost of US$100 million.

Source: Climate Leadership Group 2008b.

Box 8.5

Calgary C-Train, Alberta, Canada—Ride the Wind!

The C-Train is Calgary's light rail transit system. The system uses 39,477 MWh of electricity annually (2007 data), which the city purchases from ENMAX Energy Corporation, the city's electrical distribution provider (Inglis 2009). The program, branded as Ride the Wind!, powers the C-Train using wind energy supplied by twelve 600-kilowatt turbines. These are installed in southern Alberta on the tops of hills facing the Rockies, in order to take advantage of the strong westerly winds coming from the mountain passes. The C-Train is now 100 percent emissions-free. It is the first public light rail transit system in North America to power its train fleet with wind-generated electricity (Ride the Wind 2008).

Following capital investment of US$447 million (in the transit system and wind turbines), Calgary's C-Train saves around 590,000 t e CO_2 per year (Inglis 2009). Each day (Monday to Friday), riders board the C-Train 290,000 times. If each commuter had traveled alone in a car instead of on the C-Train, the daily mileage would have totaled 2.32 million kilometers. These car commuters would have used 238,300 liters of fuel and produced some 590,656 tonnes of carbon dioxide, as well as other pollutants such as nitrous oxide, carbon monoxide, and particulate matter.

Sources: Ride the Wind 2008; Inglis 2009.

it is clear that the strategy adopted exploits local conditions, such as high solar radiation or suitable geological conditions for geothermal energy. In other cases, the strategy was a response to local stresses, for example, traffic congestion in London and heat waves in Chicago. However, for some cases, it was just a matter of being more creative and efficient with solid waste.

Comparison with Other Datasets

The results from the G2CN case studies discussed in this chapter can be compared to those from two other datasets:

1. The Federation of Canadian Municipalities (FCM) records the expected savings in GHG emissions from some projects supported by the Green Municipal Funds. These funds, which were endowed by the Canadian government, provide grants and below-market loans to directly support municipal initiatives in Canada.
2. The United Nations Framework Convention on Climate Change (UNFCCC) has a database of projects funded under the Clean Development Mechanism (CDM). The CDM arrangement, developed under the Kyoto Protocol, allows industrialized countries to invest in emissions reduction projects in developing countries, as alternatives to more expensive strategies in their own countries.

The majority of projects in the FCM database for which both GHG savings and capital costs are reported are in the solid waste sector (12). In addition, data are available for four transportation projects, four energy supply projects, and one community development project—an eco-industrial park in Hinton, Alberta (Canada).

Generally speaking, the nine data points for the non-waste sectors are distributed in a relatively similar manner to our case study data (again, on a log-log plot). The line of best fit of G2CN data, from figure 8.2, is shown with the FCM data in figure 8.3 for comparison. The cost-effectiveness of the nine non-waste-sector projects (2,040 t e CO_2/yr/US$ million) is on average better than for the G2CN case studies (550 t e CO_2/yr/US$ million); eight of the nine points lie above the regression line of G2CN data. Furthermore, it is quite apparent that the solid waste projects in the FCM dataset substantially out-perform the data from the G2CN case studies. The average cost-effectiveness of the FCM solid waste projects is 37,400 t e CO_2/yr/US$ million.

The United Nations database of CDM projects was accessed online between January and April 2009. At that time, there were more than 1,500 registered CDM projects, with several thousand more under review. Capital cost data were taken from project design documents, assuming that capital cost was equal to incremental cost for the project (thus, operations and maintenance costs are ignored), and that the baseline scenario is to do nothing. Conversions of costs are made to U.S. dollars as of the project registration date.[6] Where capital costs could not be located in the project design document, the project was not included as a data point.

There are many solid waste and energy projects in the database. Figure 8.4 plots waste and energy projects at the high end of expected GHG reductions, that is, all energy projects over 500,000 t e CO_2/yr, and all waste projects over 250,000 t e CO_2/yr are shown, as well as a few other smaller scale projects that were of interest to the study team and some agricultural and industrial projects. There are also a small number of transportation and afforestation/reforestation projects in the CDM database; these are plotted in figure 8.4 as well.

FIGURE 8.3

Log–Log Plot of Annual GHG Savings vs. Capital Costs for Infrastructure Projects Funded Under the Federation of Canadian Municipalities' Green Municipal Fund

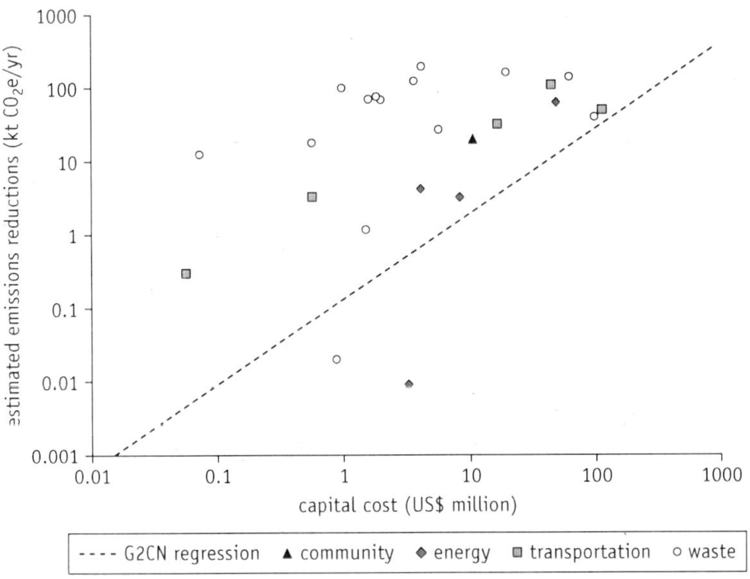

Data Sources: FCM 2009; Kennedy 2009.
Note: The dotted line is from the regression of G2CN data shown in figure 8.2.

In comparison to the G2CN data, the CDM projects are much more cost-effective, as should be expected for developing countries. All of the CDM projects lie above the regression line for the G2CN data. The spread among the CDM projects is quite remarkable. For example, the Alto-Tietê landfill gas capture project in Brazil has certified reductions of 481,000 t e CO_2/yr for a capital cost of US$2.31 million, while a bus rapid transit project in Bogota, Colombia, saves 247,000 t e CO_2/yr, at a capital cost of US$532 million. As with the FCM data, the solid waste projects are the *low hanging fruit* in terms of cost-effectiveness. This is apparent when comparing the energy and waste projects in figure 8.4 (again recognizing that these are partial data). The average cost effectiveness of the CDM waste projects is 64,800 t e CO_2/yr/US$ million, compared to 10,300 t e CO_2/yr/US$ million for energy projects.

Limitations of the Dataset

Several caveats apply to interpretation of the results above and comparison with other datasets. First, the estimation of GHG emissions for projects in the dataset used for the study has not necessarily been undertaken with a consistent methodology. Other than the few cases where the authors calculated the GHG savings, the quality

FIGURE 8.4
Log–Log Plot of Annual GHG Savings vs. Capital Costs for a Subset of United Nations Clean Development Mechanism Infrastructure Projects

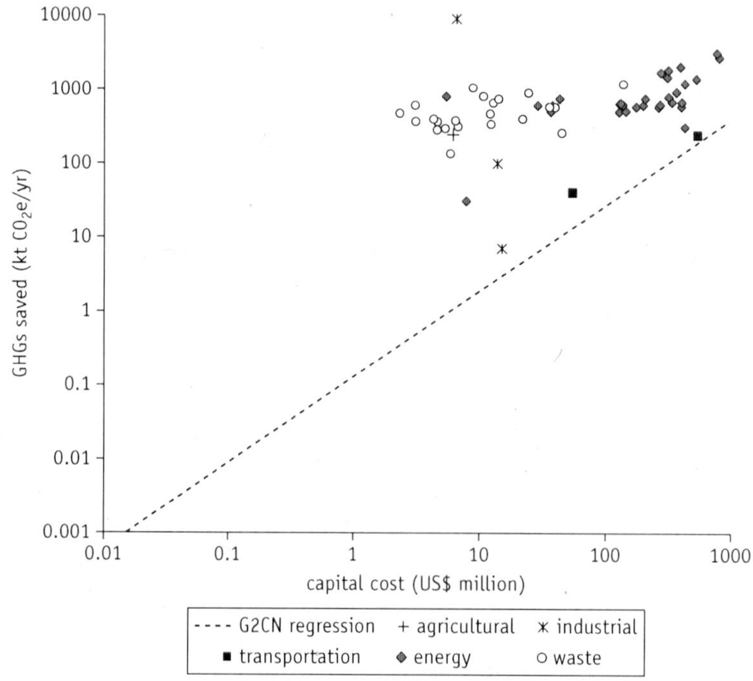

Data Sources: UNFCC 2009; Kennedy 2009.
Note: The dotted line is from the regression of G2CN data shown in figure 8.2.

of the dataset depends on the calculations undertaken individually for each project by the concerned project team.

Furthermore, the study team has undertaken a broad scan of infrastructure strategies for reducing GHG emissions. Generally, only one or two cases of a particular type of strategy are included in the dataset, and these may not necessarily be representative of the average performance of such a strategy. Where there are multiple data for a particular strategy, such as landfill–gas-to-energy in the CDM and FCM datasets, then a high degree of variation in cost-effectiveness is apparent.

Part of the variation in costs and GHG savings between projects can be attributed to differences in local conditions. Costs of projects vary because of factors such as labor costs, access to resources, access to technology, economies of scale, and so forth. GHG emissions saved when generating electricity from renewable sources depend on the GHG intensity of the local power grid. So even if costs are the same, the cost-effectiveness is higher in regions that currently have greater dependence on coal for power generation. GHG reduction strategies that are cost-effective in one region may not be so in another.

Several of the projects considered in the dataset are cutting-edge applications of new or developing technologies. As such, the costs of these projects, which may be considered trials or experiments, can be expected to come down as the technology develops.

Another important caveat is that few, if any, of the infrastructure projects considered in the dataset were designed solely for the purpose of reducing GHG emissions. Reducing emissions is only one goal. Transportation systems are designed to move people and goods; energy infrastructure is designed to provide heating, lighting, and electrical service, and so on. By virtue of differences in their functionality, various types of infrastructure can be expected to differ in terms of cost-effectiveness for reducing GHGs; this is less apparent with the G2CN data, but is clearly the case with the FCM and CDM datasets.

Finally, while cost-effectiveness has some merits as an economic measure, it is of limited use from an investment perspective. The private sector, in particular, must expect to achieve satisfactory rates of return if it is to invest in infrastructure that reduces GHG emissions. The Organisation for Economic Co-operation and Development and International Energy Agency (OECD/IEA 2008) have identified a number of energy efficiency initiatives in a few cities, such as building retrofits, LED (light emitting diode) traffic signals, and pool heat recovery, for which rates of return greater than 100 percent are achieved. Kikuchi, Bristow, and Kennedy (2009) have also shown that investments in alternative energy technologies in Ontario, Canada, can offer investors reasonable rates of return at relatively low risk, depending on the sector. The investments considered in both of the above studies, however, are relatively small in scale. Further studies of returns on investment are perhaps warranted with respect to infrastructure projects that substantially reduce GHG emissions.

Conclusions

The case studies presented in this chapter cover a variety of infrastructure strategies, with savings in GHG emissions ranging from 45 t e CO_2 to more than 500,000 t e CO_2. The size of the GHG reductions generally increases with the magnitude of capital investment, as expected, but there is significant variation among the cases. Measures of cost-effectiveness vary between 3 and 2,780 t e CO_2/yr/US$ million for the database.

The average cost-effectiveness of the G2CN database is 550 t e CO_2/yr/US$ million. This figure could be used as a benchmark to compare GHG savings for urban infrastructure investments in developed nations. However, much higher levels of cost-effectiveness have been achieved with solid waste projects in Canada (FCM 2009) and with developing world CDM projects. Thus a higher benchmark should perhaps be set for infrastructure projects in developed world cities, depending on the type of infrastructure available.

This chapter has highlighted five case studies in particular that are above the average cost-effectiveness achieved in the G2CN dataset and that have GHG savings over 100,000 t e CO_2. These cases are a solar central receiver station in Seville, Spain; the congestion charging scheme in London; combined heat and power in Gothenburg, Sweden; source separation and energy recovery in Sydney, Australia; and light rail transit in Calgary, Canada. Another four projects in the database have GHG savings over 100,000 t e CO_2, but the capital costs of these initiatives were not available to the authors.

While the GHG savings from the nine cases are substantial, these should be assessed with reference to the GHG inventories of cities. A GHG saving of 100,000 t e CO_2 per year is still two orders of magnitude below the typical emissions for a city of 1 million people, and three orders of magnitude below that for a mega-region, as discussed in the introduction and shown in table 8.1. Cities arguably need to start planning projects with reductions on the order of 1 million and greater t e CO_2 per year.

Notes

1. See International Local Government GHG Emissions Analysis Protocol Draft Release Version 1.0.: http://www.iclei.org/fileadmin/user_upload/documents/Global/Progams/GHG/LGGHGEmissionsProtocol.pdf, accessed May 2009.

2. Canadian examples include Calgary's wind-powered C-Train; Toronto's deep-lake water cooling; and sustainable neighborhood developments at Dockside Green (Victoria), South East False Creek (Vancouver), and Okotoks (near Calgary). To these, international examples, such as Malmo's port, Hammarby (Stockholm), and Kronsberg (Hannover), can be added. A few western cities, such as London, UK, and Freiburg, Germany, have reduced per capita automobile use and associated emissions. Currently under development are communities such as Masdar, near Abu Dhabi, and Dongtan, near Shanghai, China, which aim to be the world's first carbon-neutral, sustainable cities.

3. An example of a rule of thumb is that retrofitting of residential homes typically reduces average energy demand by 20 to 25 percent in Canada.

4. The authors would be pleased to receive further suggestions for infrastructure projects that substantially reduce GHG emissions; a form for submitting information on new case studies can be accessed on the project Web site: http://www.utoronto.ca/sig/g2cn.

5. For detailed case study descriptions and discussion of other benefits and implementation barriers that have been overcome, see the project Web site (http://www.utoronto.ca/sig) and the guidebook (Kennedy 2009).

6. Using the currency conversion Web site, http://www.Oanda.com, at the inter-bank rate.

References

Abengoa Solar. 2008. "PS10: The First Commercial Tower of the World." http://www.abengoasolar.com/corp/web/en/our_projects/solucar/ps10/index.html, accessed January 27, 2010.

Climate Leadership Group. 2008a. "C40 Cities Best Practices–Waste: Gothenburg, Sweden." http://www.c40cities.org/bestpractices/waste/gothenburg_system.jsp, accessed January 27, 2010.

———. 2008b. "C40 Cities Best Practices—Waste: Sydney, Australia." http://www.c40cities.org/bestpractices/waste/sydney_recycling.jsp, accessed January 27, 2010.

Codoban, N., and C. A. Kennedy. 2008. "The Metabolism of Neighbourhoods." *ASCE Journal of Urban Planning and Development* 134 (1): 21–31.

Dong, B., C. A. Kennedy, and K. Pressnail. 2005. "Comparing Life Cycle Implications of Building Retrofit and Replacement Options." *Canadian Journal for Civil Engineering* 32 (6): 1051–63.

EIA (U.S. Energy Information Administration). 2006. "State Electricity Profiles." http://www.eia.doe.gov/cneaf/electricity/st_profiles/e_profiles_sum.html, accessed January 27, 2010.

Engel-Yan, J., C. A. Kennedy, S. Saiz, and K. Pressnail. 2005. "Towards Sustainable Neighbourhoods: The Need to Consider Infrastructure Interactions." *Canadian Journal for Civil Engineering* 32 (1): 45–57.

Evans, R. 2008. "Central London Congestion Charging Scheme: Ex-post Evaluation of the Quantified Impacts of the Original Scheme." Transport for London. http://www.tfl.gov.uk/assets/downloads/Ex-post-evaluation-of-quantified-impacts-of-original-scheme-07-June.pdf, accessed January 27, 2010.

FCM (Federation of Canadian Municipalities). 2009. Green Municipal Fund Approved Projects Database. http://gmf.fcm.ca/Search/Search/Search.aspx?lang=e, accessed January 27, 2010.

IEA SolarPaces. 2008. "Spain PS10." Solar Power and Chemical Energy Systems. An Implementing Agreement of the International Energy Agency. http://www.solarpaces.org/Tasks/Task1/PS10.htm, accessed January 27, 2010.

Inglis, P. 2009. Personal correspondence with the City of Calgary.

Kennedy, C. A. 2002. "A Comparison of the Sustainability of Public and Private Transportation Systems: Study of the Greater Toronto Area." *Transportation* 29: 459–93.

———. 2007. "Applying Industrial Ecology to Design a Sustainable Built Environment: The Toronto Port Lands Challenge." Engineering Sustainability conference, Pittsburgh, PA, April 15–18.

———, ed. 2009. *Getting to Carbon Neutral: A Guide for Canadian Municipalities* Prepared for the Toronto and Region Conservation Authority. http://www.utoronto.ca/sig.

Kennedy, C. A., J. Cuddihy, and J. Engel-Yan. 2007. "The Changing Metabolism of Cities." *Journal of Industrial Ecology* 11 (2): 43–59.

Kennedy, C. A., E. Miller, A. Shalaby, H. MacLean, and J. Coleman. 2005. "The Four Pillars of Sustainable Urban Transportation." *Transport Reviews* 25 (4): 393–414.

Kennedy, C. A., and E. Mohareb. Forthcoming. "Greenhouse Gas Emissions from Cities." In *Cities in Evolution: Urbanization, Environmental Change and Sustainability*, ed. Xuemei Bai, Thomas Graedel, and Akio Morishima. Cambridge, UK: Cambridge University Press.

Kennedy, C. A., J. Steinberger, B. Gasson, T. Hillman, M. Havránek, Y. Hansen, D. Pataki, A. Phdungsilp, A. Ramaswami, G. Villalba Mendez, 2009. "Greenhouse Gas Emissions from Global Cities." *Environmental Science and Technology* 43: 7297–7302.

Kikuchi, E., D. Bristow, and C. A. Kennedy. 2009. "Evaluation of Region-specific Residential Energy Systems for GHG reductions: Case Studies in Canadian Cities." *Energy Policy* 37 (4): 1257–66.

OECD/IEA (Organisation for Economic Co-operation and Development/International Energy Agency) 2008. "Promoting Energy Efficiency, Best Practice in Cities, A Pilot Study." Working Paper, IEA, Paris.

OPA (Ontario Power Authority). 2007. "Integrated Power Systems Plan." http://www.powerauthority.on.ca/Page.asp?PageID=1224&SiteNodeID=320, accessed January 27, 2010.

Racoviceanu, A., B. W. Karney, C. A. Kennedy, and A. F. Colombo. 2007. "Life Cycle Energy Use and Greenhouse Gas Emissions Inventory for Water Treatment Systems." *ASCE Journal of Infrastructure Systems* 13 (4): 261–70.

Ride the Wind! 2008. "The C-Train Backgrounder: Ride the Wind!" Re-Energy, CA. http://www.re-energy.ca/ridethewind/backgrounder.shtml, accessed January 27, 2010.

Sahely, H. R., and C. A. Kennedy. 2007. "Integrated Systems Flow Model for Quantifying Environmental and Economic Sustainability Indicators: Case Study of the City of Toronto Urban Water System." *ASCE Journal of Water Resources Planning and Management* 133 (6): 550–59.

Sahely, H. R., S. Dudding, and C. A. Kennedy. 2003. "Estimating the Urban Metabolism of Canadian Cities: GTA Case Study." *Canadian Journal for Civil Engineering* 30: 468–83.

Saiz, S., C. A. Kennedy, B. Bass, and K. Pressnail. 2006. "Comparative Life Cycle Assessment of Standard and Green Roofs." *Environmental Science and Technology* 40 (13): 4312–16.

Transport for London. 2008. "Congestion Charging, Transport for London." http://www.tfl.gov.uk/roadusers/congestioncharging/, accessed January 27, 2010.

TRCA (Toronto and Region Conservation Authority). 2008. "Path to the Living City." Living City Project, http://www.trca.on.ca/learn/path-to-the-living-city.dot, accessed January 27, 2010.

UNFCCC (United Nations Framework Convention on Climate Change). 2009. Clean Development Mechanism database. http://cdm.unfccc.int/Projects/projsearch.html, accessed January 27, 2010.

Zachariah, L., C. A. Kennedy, and K. Pressnail. 2002. "What Makes a Building Green?" *International Journal of Environmental Technology and Management* 2(1–3): 38–53.

CHAPTER 9

Supporting Energy Efficient Solutions in Developing Countries: The Way Ahead

Ranjan K. Bose and Sangeeta Nandi

This chapter provides a contextual overview of the Energy Sector Management Assistance Program's (ESMAP) programmatic priorities to support energy efficient urban growth in developing countries; these priorities, to be effected through the Energy Efficient Cities Initiative (EECI), have been set on the basis of discussions at the fifth World Bank Urban Research Symposium and deliberations under EECI. Accordingly, the focus of the ESMAP-EECI program is (1) continued development support for analytical tools and assessment approaches with regard to new cities, expanding cities, and retrofitting in urban sectors; (2) establishment of operationally relevant urban data systems that support analytical tools for citywide and sector-specific energy interventions, as well as the assessment and benchmarking of good practices; (3) exchange of knowledge products and good practices to influence better-informed policy making by city authorities for concrete action on energy efficient initiatives; and (4) building institutional partnerships to advance the development of holistic approaches on urban energy efficiency. In addition, ESMAP also directly supports the development of an analytical tool, the Rapid Assessment Framework (RAF), for the energy efficient retrofitting of city sectors. Expected to be ready for implementation by the EECI by September/October 2010, RAF will complement ongoing World Bank policy dialogue, sector work, partnerships, and investment opportunities, as well as efforts by partner organizations to facilitate energy and climate solutions in existing cities.

Instituting Energy Efficiency and Climate Resilience in Developing Countries

The ramifications of current trajectories of energy intensive urbanization extend beyond national energy security concerns to encompass associated increases in

Ranjan K. Bose, PhD, is Senior Energy Specialist at the World Bank; Sangeeta Nandi, PhD, is an Independent Consultant who specializes in sustainable economic development issues.

global-climate-threatening green house gas (GHG) emissions. The consequent climate risks the development and mainstreaming of energy efficient and low-carbon pathways that facilitate urban progress in an environmentally sustainable manner. The implementation of climate-friendly urban energy solutions is made more urgent by the fact that most rapid urbanization is taking place in the developing regions of Africa and Asia, which are least able to cope with the uncertainties and extremities of climate impacts. Against this backdrop, this volume is a timely contribution to the body of knowledge on urban energy efficiency. Its chapters, which were presented at the World Bank's fifth Urban Research Symposium, "Cities and Climate Change: Responding to an Urgent Agenda," in June 2009, put into perspective the Energy Sector Management Assistance Program's (ESMAP) prioritization of citywide and sector-specific urban energy initiatives with regard to (1) tools and assessment approaches on energy efficient urban development and (2) good practices that promote low-carbon interventions in urban sectors.

The analytical tools and policy insights discussed in this volume extend from integrated assessments of new cities to the impacts of socioeconomic, climatic, and demographic changes on existing cities. Sector-specific interventions are discussed in the context of tools to "green" buildings in Australia, the transformation to efficient lighting systems in the Philippines, and demand responsive transport (DRT) systems in France. In addition, the documentation and benchmarking of a variety of low-carbon and carbon-neutral good practices provides a range of practical insights on plausible energy efficient interventions in urban sectors. Thus, the chapters in this publication comprise significant contributions to the ESMAP objective of mainstreaming and leveraging knowledge and initiatives on urban energy efficiency. Following from them, this chapter provides a contextual overview of ESMAP's programmatic priorities to support energy efficient urban growth, which are to be effected through the Energy Efficient Cities Initiative (EECI). On the basis of discussions at the Urban Research Symposium and deliberations under EECI, these priorities were set out as follows:

- Continued development support for analytical tools and assessment approaches with regard to new cities, expanding cities, and retrofitting in urban sectors.
- Establishment of operationally relevant urban data systems that support analytical tools for citywide and sector-specific energy interventions, as well the assessment and benchmarking of good practices.
- Exchange of knowledge products and good practices to influence better-informed policy making by city authorities for concrete action on energy efficient initiatives.
- Building institutional partnerships to advance the development of holistic approaches on urban energy efficiency.

The above priorities are in consonance with ESMAP's central approach of facilitating policy-relevant knowledge development and dissemination for the mainstreaming of energy efficient initiatives, to be implemented in the urban context through EECI. ESMAP also directly supports the development of an analytical tool, the Rapid Assessment Framework (RAF), for the energy efficient retrofitting of city sectors. Expected to be ready for implementation by the EECI by September/October 2010, RAF will complement ongoing World Bank policy dialogue, sector work, partnerships, and investment opportunities, as well as efforts by partner organizations to facilitate energy and climate solutions in existing cities.

A Rapid Assessment Framework for Energy Efficient Cities: A Programmatic Approach

The body of knowledge on analytical tools and assessment approaches on urban energy use and climate impacts, to which chapters 2, 3, and 4 are significant contributions, largely focuses on new city developments and extensions to existing cities as a result of socioeconomic and demographic changes. In addition, there is also a great potential to realize energy savings in cities through energy efficient retrofits of existing infrastructure systems and buildings. For example, the continued use of outdated and obsolete equipment to provide basic municipal services remains a key challenge in almost all developing country cities. To address this problem, ESMAP has launched the RAF. RAF is being developed as a practical tool for conducting rapid assessments in cities to identify and prioritize retrofitting opportunities for energy efficiency interventions and to suggest intervention options. When ready for implementation, RAF would focus on energy use in six city sectors: transport, buildings, water, public lighting, heating and power, and solid waste. Designed to provide a quick analysis of systems characterized by financial and informational constraints, as well as socioeconomic imperatives, RAF will evaluate sectors based on costs, energy use, and GHG profiles. Additionally, the framework will have the capacity to quantify economic and environmental co-benefits of energy efficient interventions. Box 9.1 provides a snapshot view of the structure of RAF, whose modules are currently under development. It is envisaged that RAF will become a global framework that will allow for cross-city comparisons and the sharing of good practices on energy-saving initiatives in the six city sectors mentioned above. However, the extent of this effort will be largely determined by the availability of relevant information to support the constituent tools of RAF, thus reinforcing the importance of structured data systems.

Operationally Relevant Databases: Necessary Support Tools for Urban Energy Efficiency

Consolidated data and information systems are of vital importance to the systematic analysis and implementation of energy interventions in urban sectors. In most developing countries, inadequacies in data systems hamper the implementation of analytical tools and assessment approaches that are being developed, for example, those presented in the fifth Urban Research Symposium and RAF. These data inadequacies

Box 9.1

The Rapid Assessment Framework: A Practical Tool for Instituting Urban Energy Efficiency

The RAF, designed to present a quick, first-cut, sectoral analysis on energy use, will be developed as a simple, low-cost, user-friendly, and practical tool that can be applied in any socioeconomic setting. The framework will comprise the following:

- an Analytical Tool to assess a city's current energy efficiency profile.
- an Energy Efficiency Options Tool to help city administrations (CAs) decide what actions to take to improve efficiency.
- a Good Practice Tool that provides a structured selection of sector-specific as well as citywide examples of successful energy efficiency strategies.
- a Matrix Tool for a CA to assess its own institutional capabilities to form and implement energy efficiency policy and programs.

In the RAF structure of implementation, an initial assessment of the strengths and weaknesses of existing energy efficiency inventories and tools will be followed by the development of the Analytical Tool. The output will provide CAs with a graphical, sector-wise summary of performance and compare it with a range of peer cities. Hence, each CA will be able to quickly determine the sectors that offer the best potential for deriving energy efficiency gains. This process can be subjected to further interpretation by presenting potential energy efficiency gains in monetary values, carbon emissions reductions, or a range of other benefits such as social and environmental considerations.

The Energy Efficiency Options Tool will be used after the Analytical Tool to develop and appraise options for improving energy efficiency. This tool will take the form of a series of modules that will help the CA consider the reasons behind poor energy performance and identify the projects and programs that will help them make efficiency gains. The output of this process will be a set of "quick win" initiatives for each sector and more detailed sectoral strategies. This will enable the selection of the most effective energy efficiency options, based on a set of appraisal criteria.

Source: ESMAP documentation.

also tend to inhibit the emergence of rigorous energy and climate assessments based on past trajectories of energy use in developing countries that can subsequently be empirically verified. Therefore, it is important that support be extended to client countries to collate and consolidate their information systems such that operationally relevant databases can support the application of urban energy efficiency tools.

The availability of adequate and relevant information systems across city sectors is also important for both sector-specific urban energy interventions and the benchmarking of good practices. This is demonstrated by the sector studies presented in chapters 5, 6, and 7, and the compilation of good practices in chapter 8. For better-informed policy making, data systems used to formulate and implement decision-support tools need to be structured in a manner that lends itself to analysis. They should also include comprehensive information on costs, socioeconomic variables, environmental implications, and demographic change. This would further the advancement of energy efficient urban solutions through the documentation and dissemination of good practices that can be adapted and scaled to meet locale-specific requirements.

Dissemination and Outreach

ESMAP is in the process of documenting good practices on urban energy use from across the world in an effort to promote the sharing of successful urban energy interventions and thus to encourage the further development of strategies adapted to local specificities. Chapter 8, which documents and benchmarks a set of low-carbon urban infrastructure strategies, as well as the sector-initiatives in chapters 5, 6, and 7, contribute to this endeavour. Also, an important component of the flagship ESMAP energy efficiency diagnostic tool, RAF, described above, requires the structured selection of citywide and sector-specific good practices. A structured documentation process can provide valuable policy-insights for the determination of future approaches on urban energy efficiency by city authorities.

ESMAP has established an online database to share good practices from city sectors around the world to promote the widespread dissemination and outreach of urban energy efficiency initiatives. To populate this database, the EECI team appraises urban energy efficiency initiatives on a continual basis according to a structured format before uploading them onto the ESMAP Web site.[1] ESMAP will also institute a good practices award that recognizes innovation in city energy programs to complement the dissemination of urban energy good practices through the online database. To select awardees, ESMAP will invite cities to submit proposals periodically according to a standard proposal form, which would require comprehensive economic and technical data. The compilation of a diverse portfolio of good practices would also provide ESMAP with a practical tool for identifying emerging institutional models and instruments on urban energy initiatives and scale up financing for them accordingly.

Mobilizing Partners and Financing for Energy Efficient Urbanization

Given the complexities inherent in the cross-cutting nature of energy interventions and the enormity of the task at hand, forging strong and strategic partnerships becomes an important component of ESMAP's programmatic support for the implementation of energy efficiency in urban sectors. In a broad approach adopted to foster these partnerships, ESMAP, through EECI, works with global partner institutions to ensure that cities have access to the skills and comparative advantages of different organizations in a more holistic way. Such a partnership is already under way between EECI and Cities Alliance, a World Bank–managed program on city development and slum upgrading, to enable proactive steps for fostering vastly improved access to knowledge, policies, and technologies about energy provision and increased energy efficiency. These efforts will focus at the household, city, and national levels. Under the partnership, new proposals submitted to Cities Alliance will be offered the opportunity to access the skills, resources, and networks of ESMAP. In addition, Cities Alliance and ESMAP have agreed to jointly commission and finance a detailed policy analysis of the main energy issues that need to be addressed in an urbanizing world, particularly in the slums of developing country cities. This work will allow both organizations to better frame the partnership. As a first step, Cities Alliance will collaborate with ESMAP to provide inputs to its Global Energy Assessment.

In addition to these alliances, a proposed ESMAP and IBNET[2] (International Benchmarking Network for Water and Sanitation Utilities) partnership will establish a comprehensive energy module in the existing IBNET tool kits for the water and sanitation sector. By providing a common set of data definitions, cost indicators, performance indicators, and resources for the compilation, analysis, and presentation of energy use data, this effort would enable water utilities to set up the baseline for energy audits and monitor performance for energy efficiency improvements. EECI has also been assigned as liaison with the World Bank Group operational units to mobilize local funds, Global Environment Facility grants, Clean Technology Fund resources, carbon finance, and other financing as required to facilitate a greater scale of urban energy efficiency interventions.

Inherent Complexities in Instituting Urban Energy Efficiency

The increasing global climate footprint of energy intensive urban growth makes imperative the development of efficient, low-carbon and carbon-neutral urban energy solutions. These solutions can be implemented for entire cities or through sector-specific interventions. However, short-term energy delivery solutions are often given priority over more sustainable, climate-friendly, energy efficiency investments in urban areas. This is especially true in developing countries where infrastructure provisions and services typically lag far behind demographic growth and geographic expansion of urban settlements. This shortcoming in policy making is compounded by the cross-cutting nature of energy services, which often leads to energy efficiency decisions regarding their efficient use falling through the cracks of administrative silos. Thus, there is an urgent need to develop decision support tools to leverage urban energy efficiency interventions. These tools must account for various socioeconomic compulsions and infrastructure requirements, as well as administrative complexities, in a rapidly urbanizing world. In developing counties, this decision making should factor in the scale and nature of resource constraints and the low coping capacities of a large number of urban residents to systemic inadequacies and climate risks. The energy efficient urban development challenge thus requires the development and documentation of well-calibrated tools and assessment approaches to urban energy efficiency, as well as energy efficient sector interventions and good practices. This is central to the urban agenda of ESMAP, and is being implemented through EECI in concert with partners from client cities as well as the global development community.

Notes

1. These good practices can be found at http://www.esmap.org

2. IBNET is sponsored by the UK Department for International Development under its Water and Sanitation Programme. IBNET's objective is to support access to comparative information that will promote good practices among water supply and sanitation suppliers worldwide and eventually provide consumers with access to efficient, high-quality, affordable water supply and sanitation services.

Index

Boxes, figures, notes, and tables are indicated by *b, f, n,* and *t,* respectively.

G

Z

Appendix

FIGURE 2.2
Layout of the Eco-Town as Planned

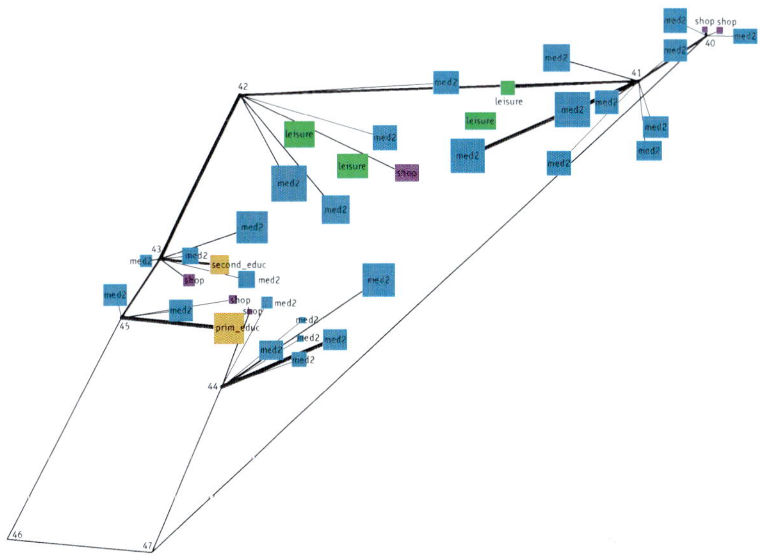

Source: Authors.

FIGURE 2.3
Cost-Optimized Eco-Town Layout (unconstrained)

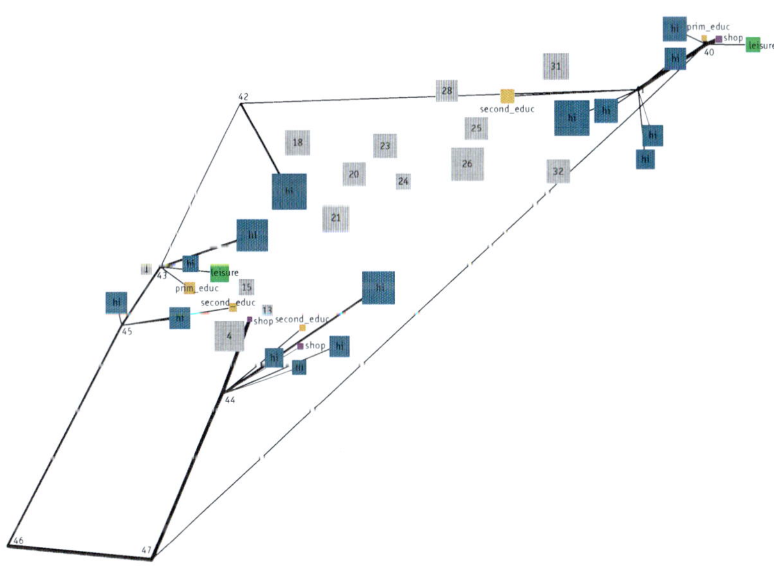

Source: Authors.

FIGURE 2.4

Cost-Optimized Eco-Town Layout (with additional planning constraints)

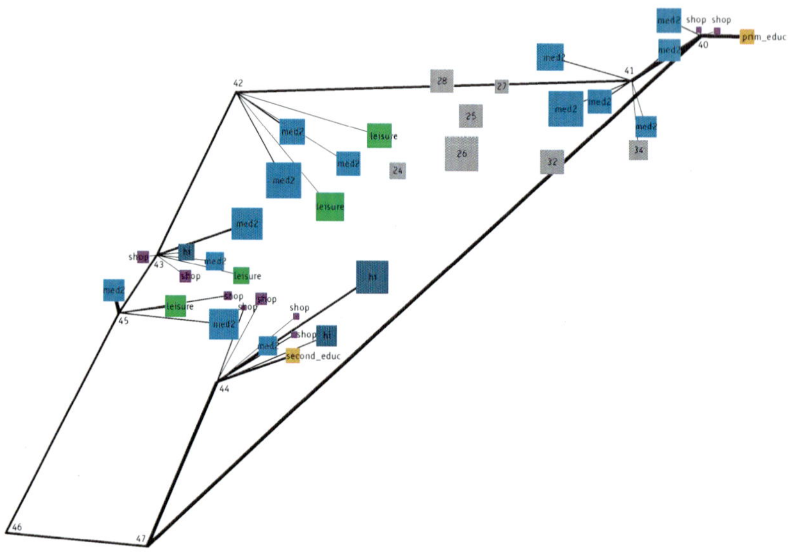

Source: Authors.

FIGURE 3.2

Zones of Development in London and the Thames Gateway

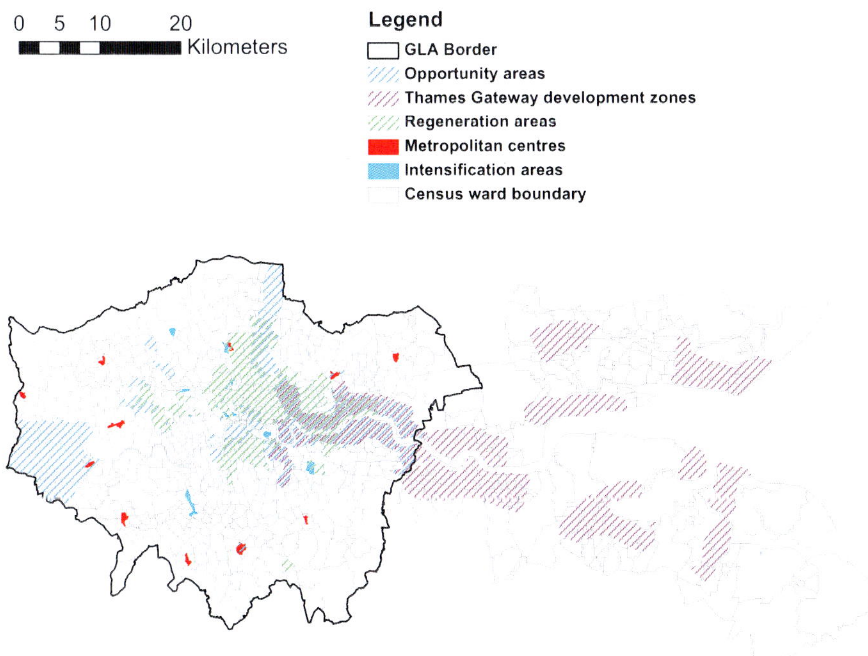

Source: Zones of development, opportunity areas, urban green space, and previously developed land within the GLA and Thames Gateway are from the Greater London Authority, http://www.london.gov.uk/thelondonplan/.

FIGURE 3.4

Example of Generalized Travel Costs by Car (in minutes) from Heathrow Ward

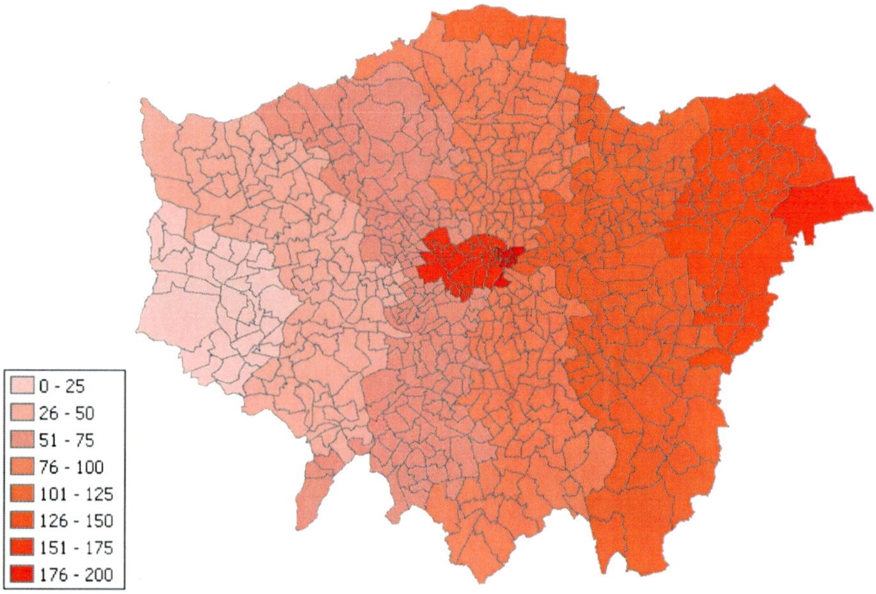

Sources: Road network. OS Mastermap ITN Data, http://www.ordnancesurvey.co.uk/oswebsite/products/osmastermap/layers/itn/; the London Travel Report, Transport for London, http://www.tfl.gov.uk/corporate/about-tfl/publications/1482.aspx.

Note: Note the influence of the congestion charging zone in central London on travel cost.

FIGURE 3.5

Projected Population Change to 2100 at a Ward Scale (high economic growth, unconstrained development)

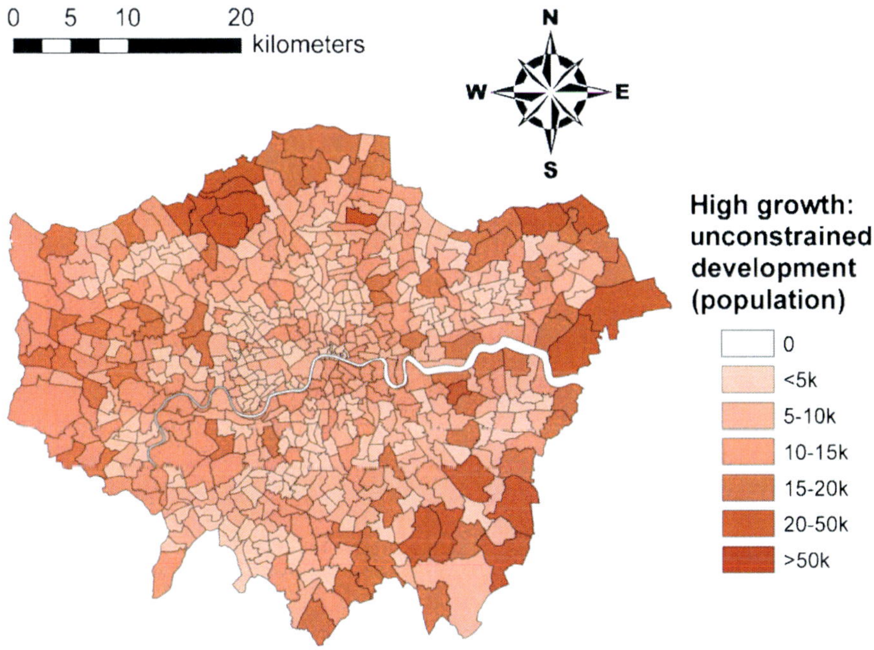

Sources: Trend-based employment forecasts for London by borough, Greater London Authority, http://www.london.gov.uk/mayor/economic_unit/docs/ep-technical-paper-1.pdf; current and historic population demographics, Office for National Statistics, http://www.statistics.gov.uk/default.asp; wards: census area statistics, U.K. Borders, http://edina.ac.uk/ukborders/; census data, U.K. Census Service, www.census.ac.uk; current land development, MasterMap, http://www.ordnancesurvey.co.uk/oswebsite/products/osmastermap/; index of deprivation, communities and local government, http://www.communities.gov.uk/communities/neighbourhoodrenewal/deprivation/deprivation07/; and property type and location, National Property Database, The Environment Agency, http://www.environment-agency.gov.uk.

FIGURE 3.6

Projected Population Change to 2100 at a Ward Scale (high economic growth, constrained development)

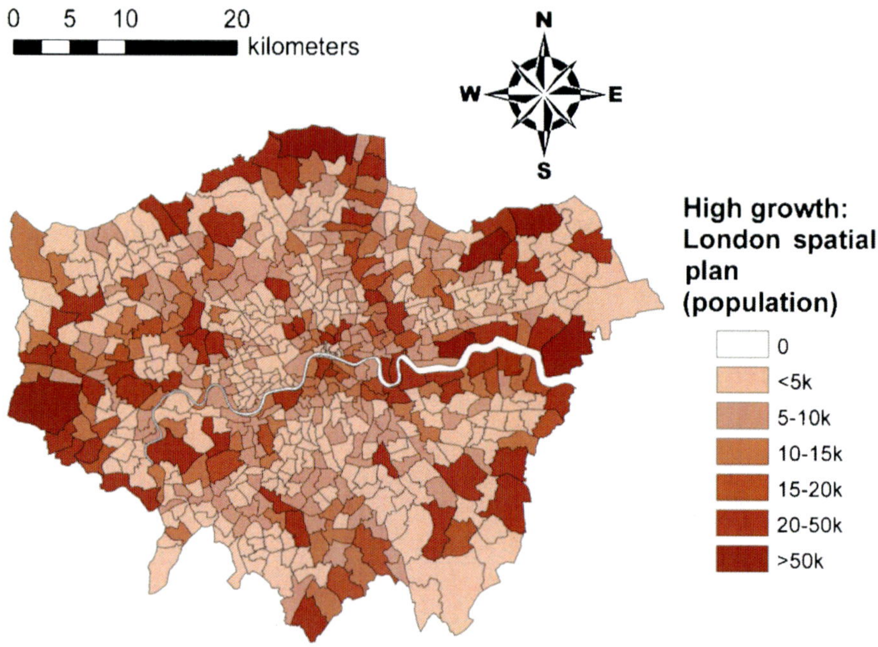

Sources: Trend-based employment forecasts for London by borough, Greater London Authority, http://www.london.gov.uk/mayor/economic_unit/docs/ep-technical-paper-1.pdf; current and historic population demographics, Office for National Statistics, http://www.statistics.gov.uk/default.asp; wards: census area statistics, U.K. Borders, http://edina.ac.uk/ukborders/; census data, U.K. Census Service, www.census.ac.uk; current land development, MasterMap, http://www.ordnancesurvey.co.uk/oswebsite/products/osmastermap/; index of deprivation, communities and local government, http://www.communities.gov.uk/communities/neighbourhoodrenewal/deprivation/deprivation07/; property type and location, National Property Database, The Environment Agency, http://www.environment-agency.gov.uk.

FIGURE 3.8

Daily Maximum Summer Temperature for London, 1961–90 and 2050

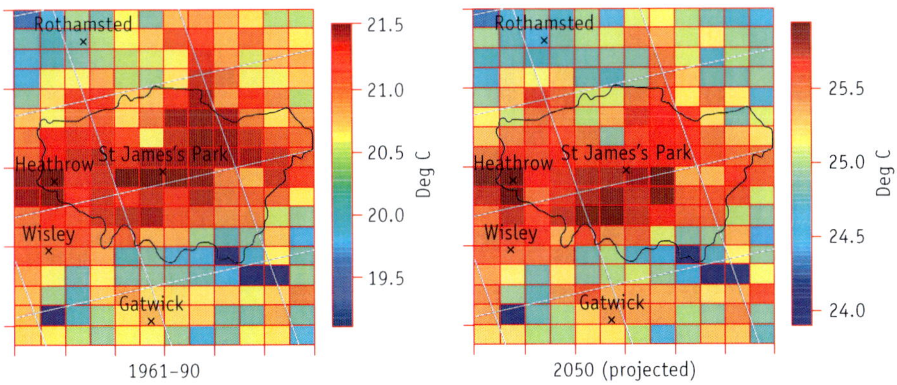

Source: Climate change projections, U.K. Climate Projections (UKCP09), http://ukcp09.defra.gov.uk/.
Note: Daily maximum summer temperatures, averaged over June, July, and August.

FIGURE 3.11

Screen Shot from the User Interface of the Assessment Tool

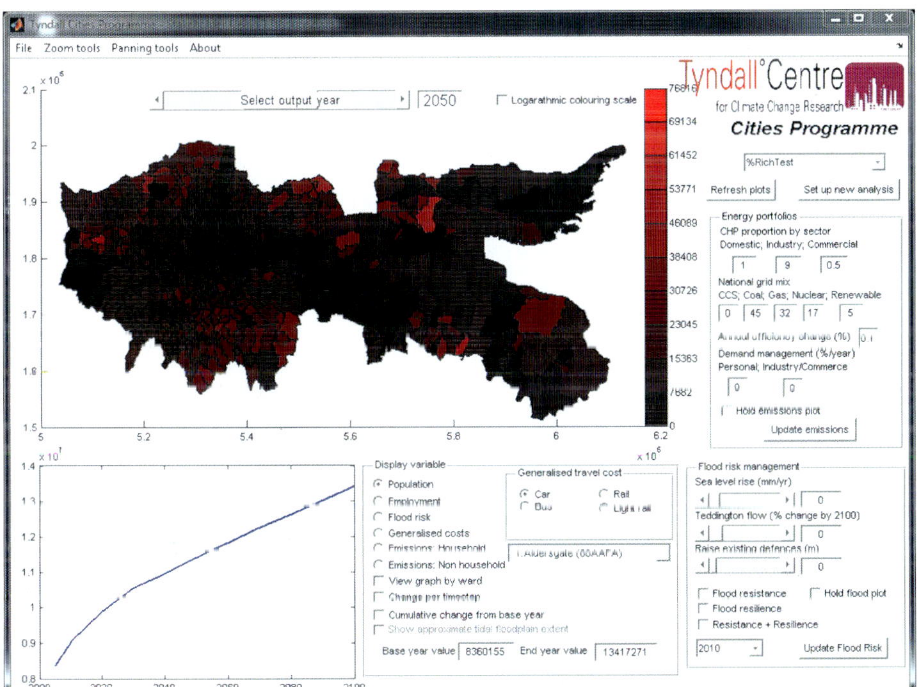

Source: Tyndall Cities Programme, Tyndall Centre for Climate Change Research.

FIGURE 4.2
Sydney in the Context of the Metropolitan Strategy

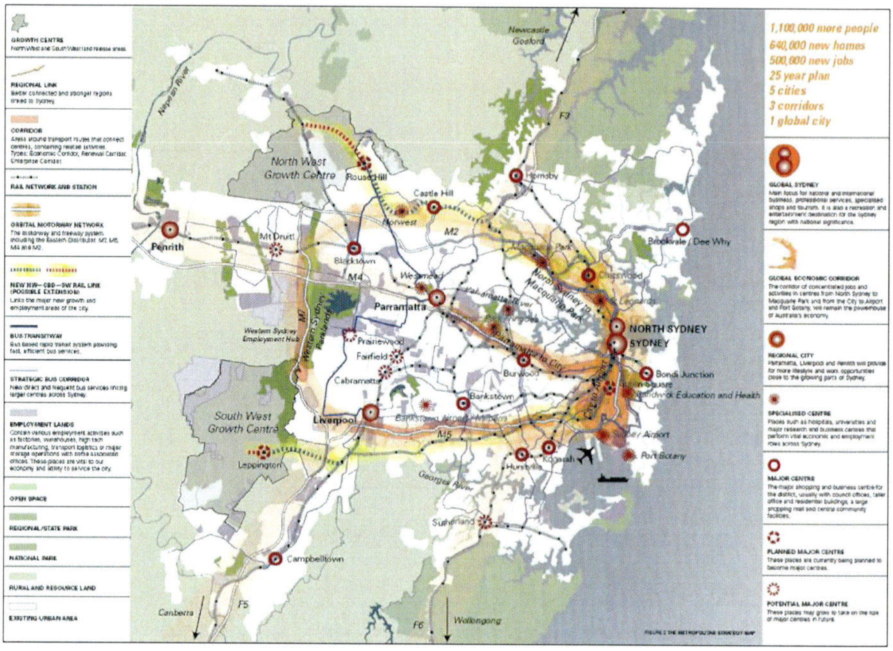

Source: NSW Department of Planning 2005, 10–11. Reproduced with kind permission of New South Wales Government Department of Planning.

FIGURE 4.3

Sydney Exogenous Housing Inputs: New Dwellings per Hectare, 2006–31

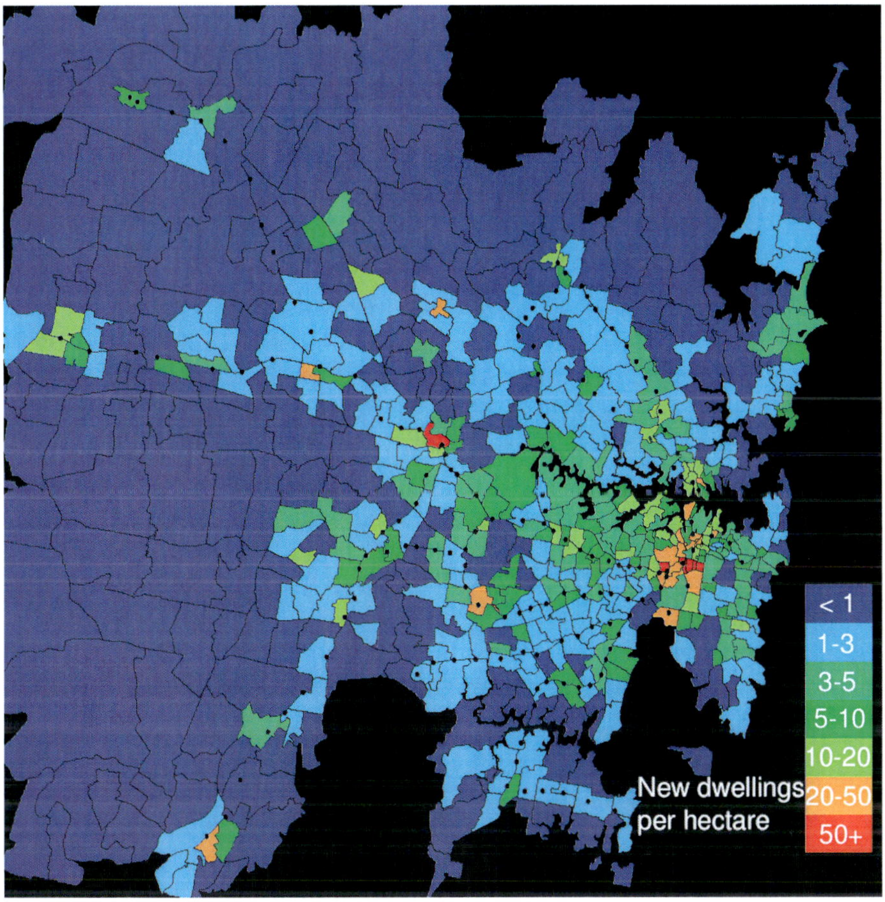

Source: Authors, based on Rickwood 2009a, 252.

FIGURE 4.4

Sydney Household Income Deciles in 2001 and 2031

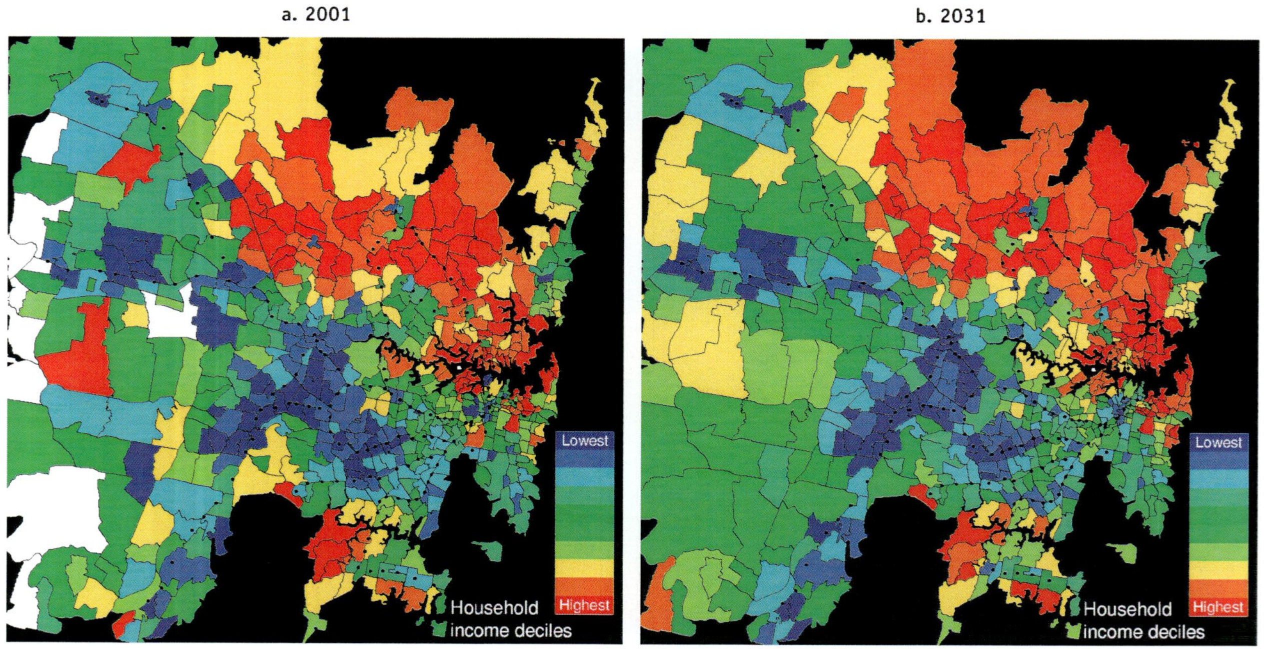

Sources: Authors, based on ABS 2001, 2006; Rickwood 2009a, 271.

Note: Household income deciles in (a) 2001 from ABS data and (b) 2031 projected for baseline scenario.

FIGURE 4.5
Sydney per Capita Income Deciles in 2031

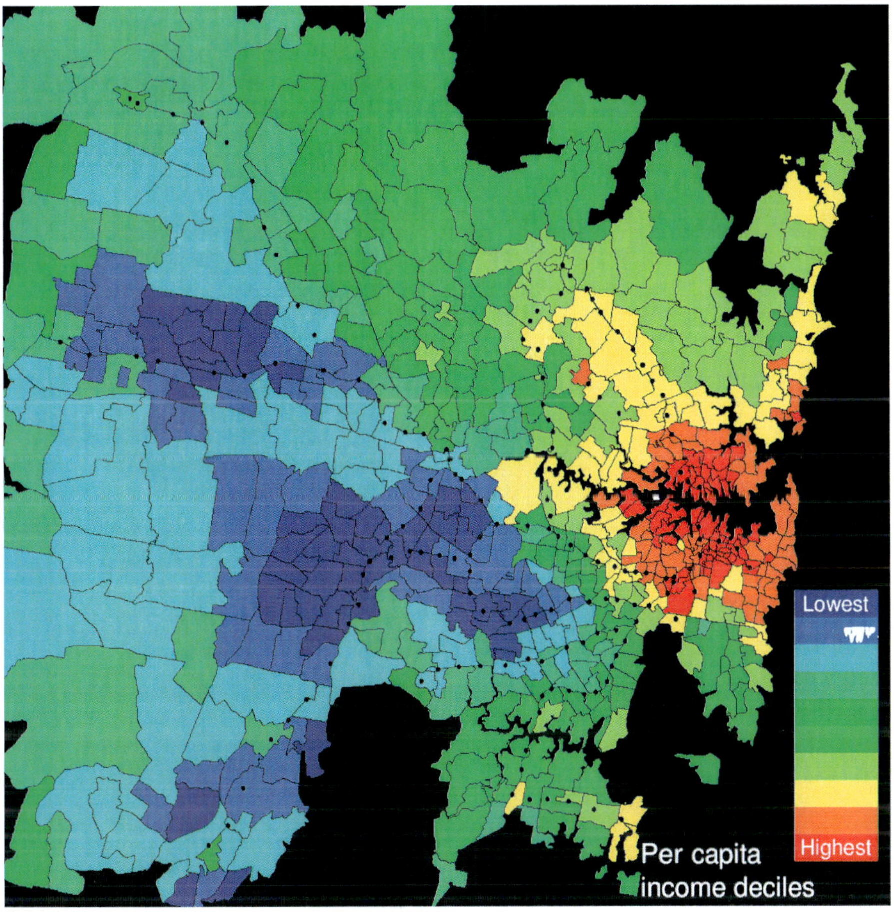

Source: Authors, based on ABS 2006 data.
Note: Per capita income deciles in 2031 (projected for baseline scenario).

FIGURE 4.6

Sydney Annual Dwelling-Related Energy Use by Zone in 2031

a. Per household

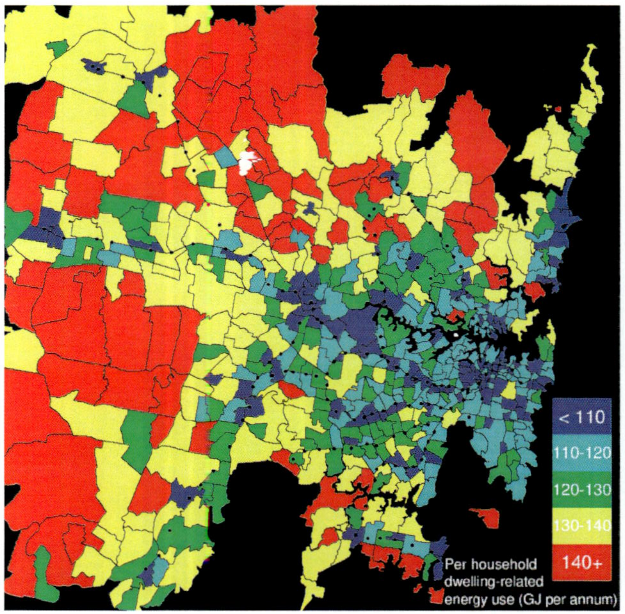

b. Per person

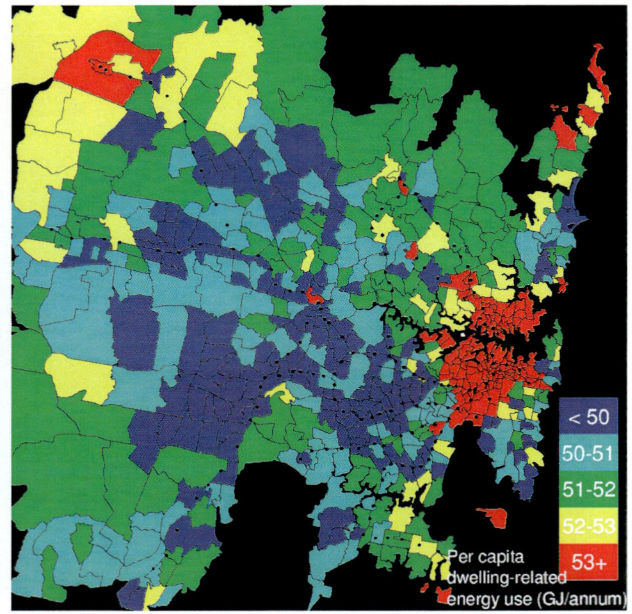

Sources: Authors, based on ABS 2001, 2006; IPART 2004a, 2006; Rickwood 2009a, 2009b.
Note: Includes embodied energy.

FIGURE 4.7

Sydney Annual Personal-Transport-Related Energy Use by Zone in 2031

a. Per household b. Per person

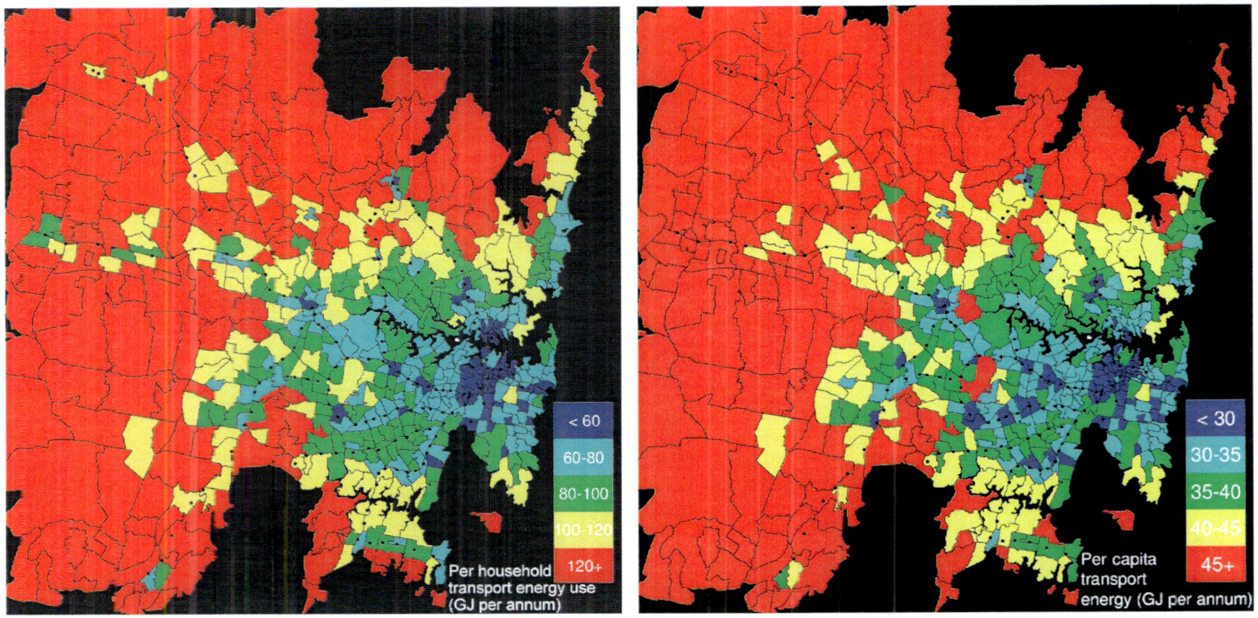

Sources: Authors, based on ABS 2001, 2006; Rickwood 2009a; TDC 2007.

Note: Includes energy embodied in cars.

FIGURE 4.8

Sydney Annual Dwelling-Related Emissions per Person by Zone in 2031

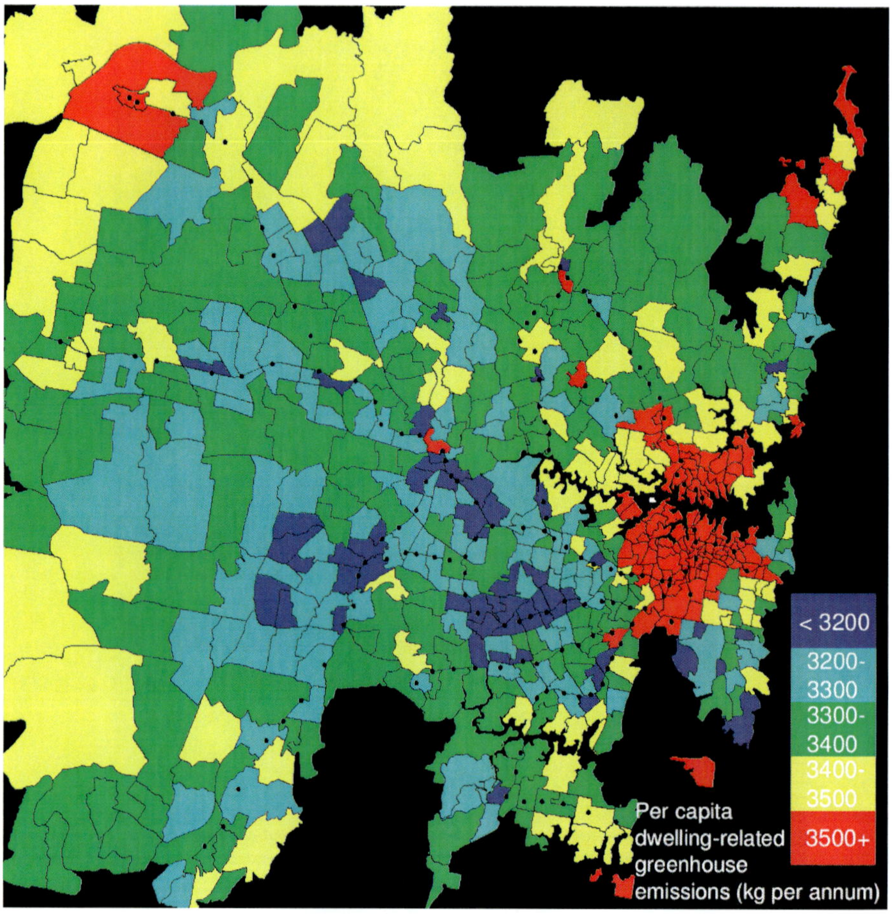

Sources: Authors, based on ABS 2001, 2006; IPART 2004a, 2006; Rickwood 2009a, 2009b.
Note: Annual dwelling-related emissions per person includes embodied energy.

FIGURE 4.9

Sydney Annual Personal-Transport-Related Emissions per Person by Zone in 2031

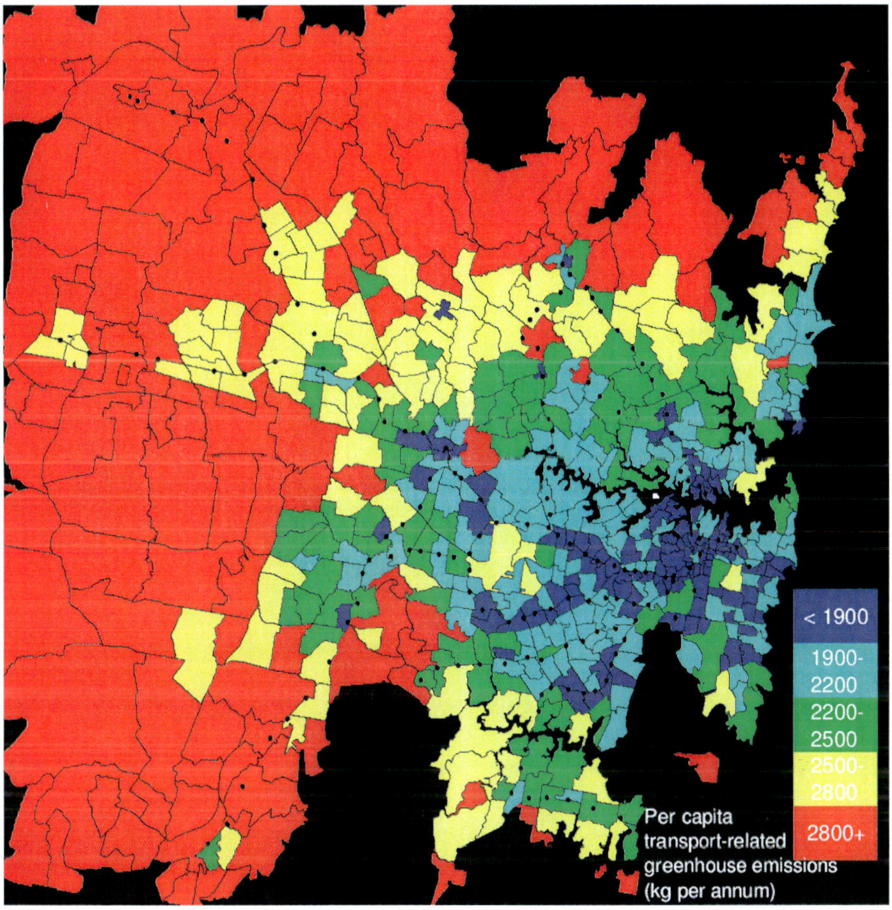

Sources: Authors, based on ABS 2001, 2006; Rickwood 2009a; TDC 2007.
Note: Includes emissions embodied in cars.

FIGURE 4.10

Sydney Annual Emissions per Person by Zone in 2031

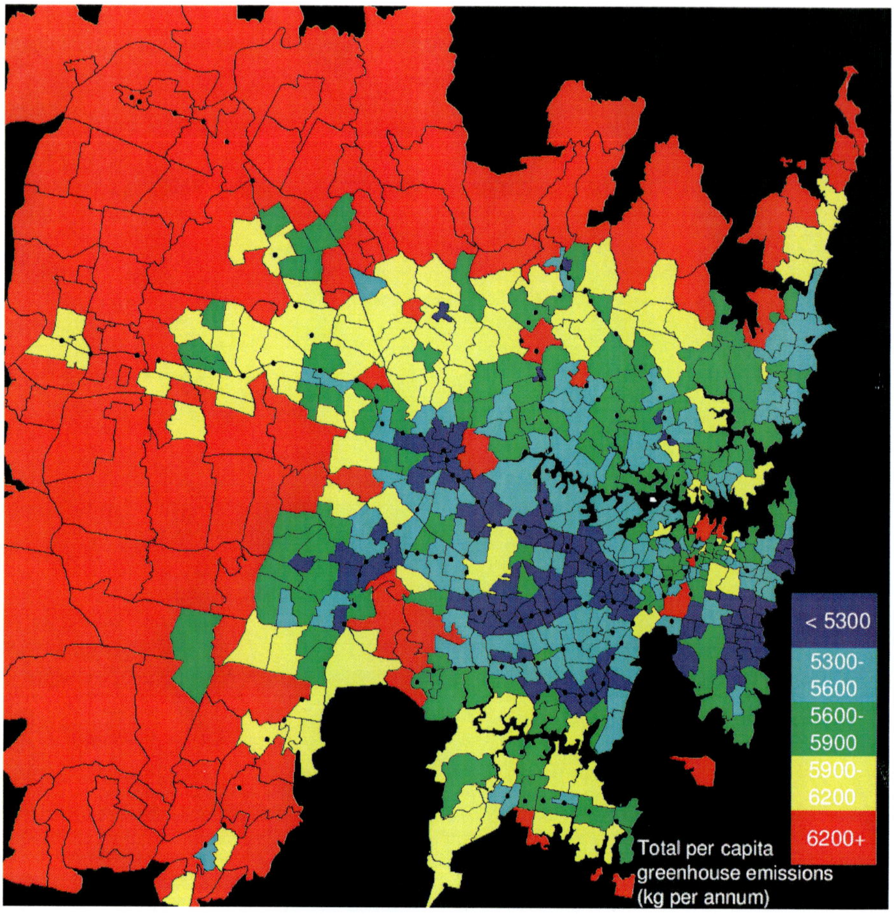

Sources: Authors, based on ABS 2001, 2006; IPART 2004a, 2006; Rickwood 2009a; TDC 2007.
Note: Includes emissions embodied in cars.

FIGURE 6.2

Market Structure with PELMATP

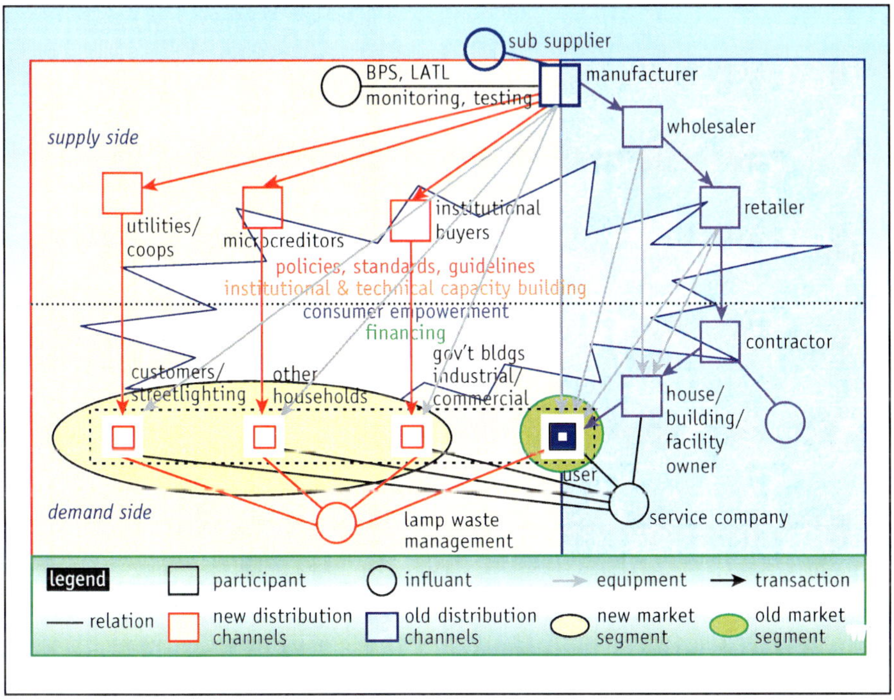

Source: Philippines DOE and UNDP 2002.

FIGURE 7.2
DRT Types and Locations in France, 2007

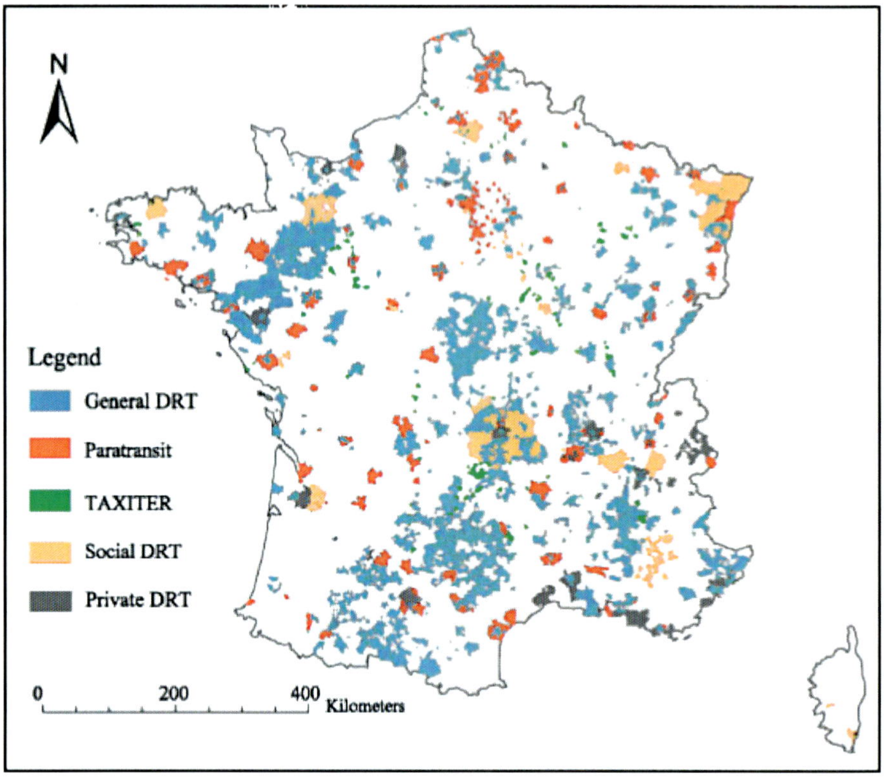

Source: Castex 2007.

FIGURE 7.6

Trip Optimization Using GaleopSys Software (TADOU)

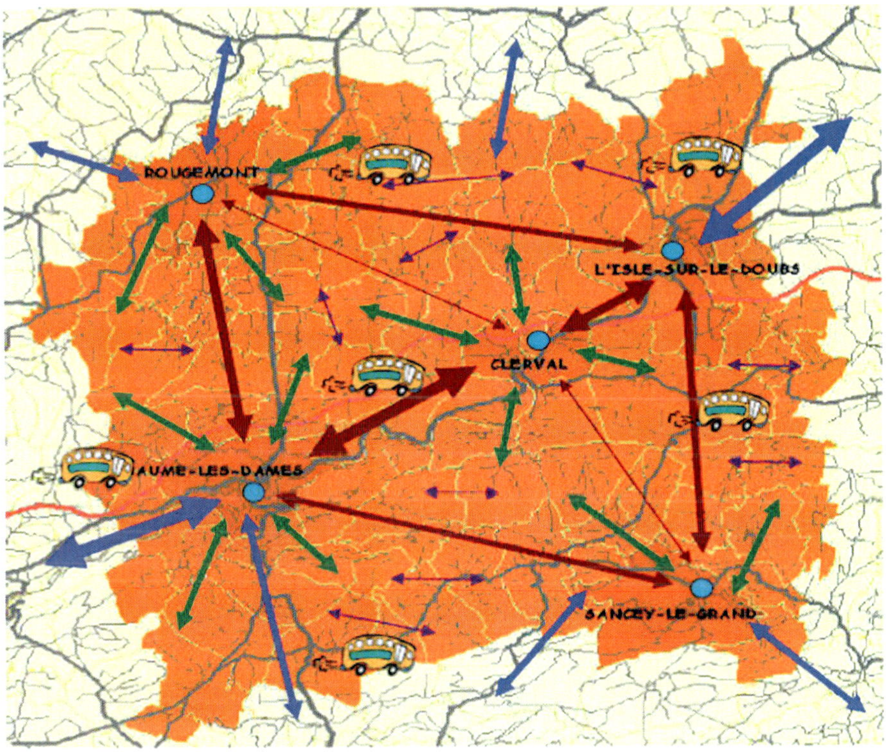

Source: Castex 2007.

FIGURE 7.7

Principles of DRT TAD 106 in Toulouse

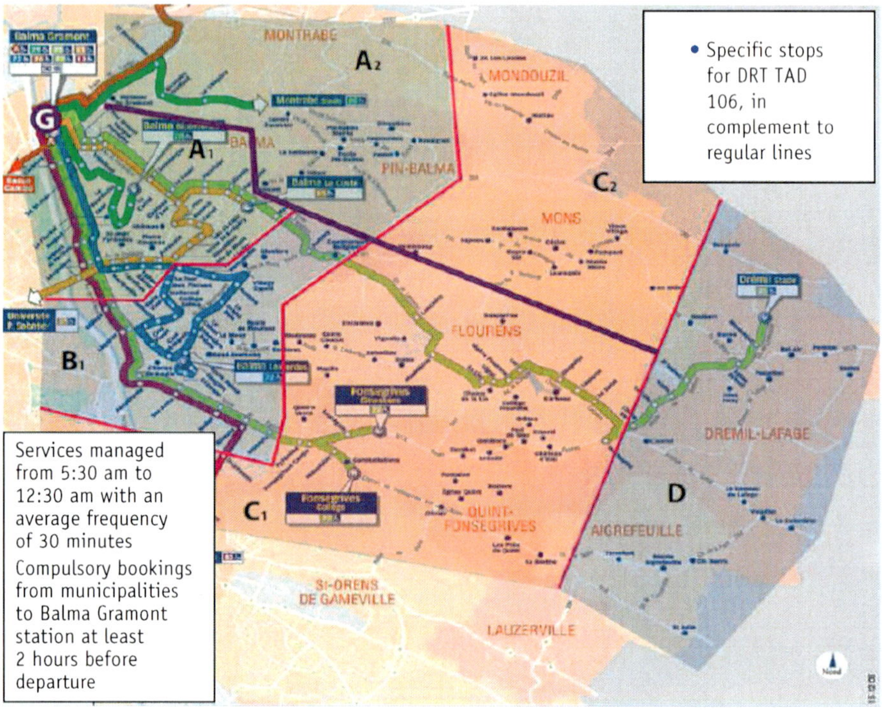

Source: DRT operator in Toulouse.

FIGURE 8.1

Locations of Case Studies for the Getting to Carbon Neutral Project

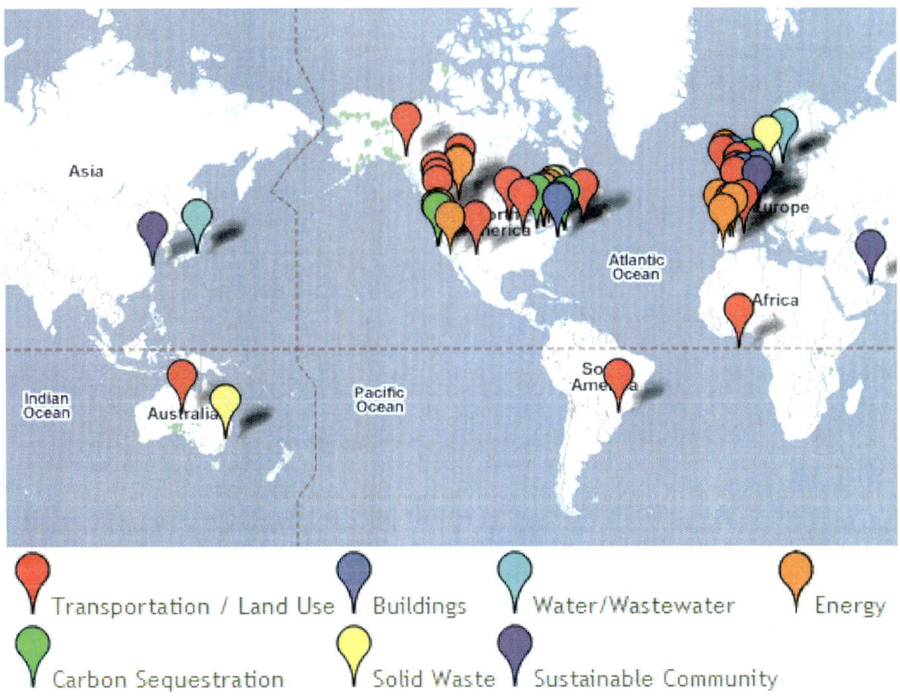

Source: Google map adapted from the project Web site: http://www.utoronto.ca/sig.

ECO-AUDIT

Environmental Benefits Statement

The World Bank is committed to preserving endangered forests and natural resources. The Office of the Publisher has chosen to print *Energy Efficient Cities* on recycled paper with 50 percent postconsumer fiber in accordance with the recommended standards for paper usage set by the Green Press Initiative, a non-profit program supporting publishers in using fiber that is not sourced from endangered forests. For more information, visit www.greenpressinitiative.org.

Saved:

- 10 trees
- 3 million Btu of total energy
- 978 lb. of net greenhouse gases
- 4,709 gal. of waste water
- 286 lb. of solid waste

green press INITIATIVE